American Zoos During the Depression

To Ken Kawata

American Zoos During the Depression

A New Deal for Animals

Jesse C. Donahue and
Erik K. Trump

McFarland & Company, Inc., Publishers
Jefferson, North Carolina, and London

LIBRARY OF CONGRESS CATALOGUING-IN-PUBLICATION DATA

Donahue, Jesse.
American zoos during the depression : a new deal for animals / by Jesse C. Donahue and Erik K. Trump.
p. cm.
Includes bibliographical references and index.

ISBN: 978-0-7864-4963-7
softcover : 50# alkaline paper ∞

1. Zoos — United States — History. 2. United States — Social conditions — 1918–1945. 3. Depressions — 1929 — United States. I. Trump, Erik. II. Title.
QL76.5.U6D657 2010
590.7'3730904 — dc22 2010030885

British Library cataloguing data are available

Cover image: detail of a Federal Art Project poster for the Brookfield Zoo, 1936 (Library of Congress)

Manufactured in the United States of America

McFarland & Company, Inc., Publishers
Box 611, Jefferson, North Carolina 28640
www.mcfarlandpub.com

Contents

Acknowledgments

We owe many people thanks for their support and assistance during the writing of this book. First among them we wish to thank Ruth and Ted Braun for the generous grant that allowed one of us to take three summers away from teaching to work on this history. Travel grants provided by Saginaw Valley State University (SVSU) funded our visits to several of the archives that were crucial for our study.

In our archival research we were assisted greatly (and kindly) by many people, including Ellen Alers at the Smithsonian Institution Archives; Anne Frantilla, Assistant City Archivist, Seattle Municipal Archives; Dana Payne, Reptile Curator at the Woodland Park Zoo; Sybil Boudreaux at the Earl K. Long Library, University of New Orleans; Irene Wainwright at the City Archives of the New Orleans Public Library; John Slate at the Dallas Municipal Archives; Steve Johnson, formerly with the Wildlife Conservation Society Archives; Joy Weiner, archives specialist at the Smithsonian Institution Archives of American Art, New York City; Kenneth Cobb and Leonora Gidlund at the City of New York's Municipal Archives; Scott Sendrow, Parks Librarian with the New York City Department of Parks and Recreation Library; Patricia Salmon, at the Library of the Staten Island Institute of Arts and Sciences; John Caltabiano, Ellen Palm, and Matt Lanier of the Staten Island Zoo; Laura Ruttum at the New York Public Library, Manuscripts and Archives Division; Amy Peck at the Prospect Park Archives; and Doris Wesley and Kenn Thomas at the Western Historical Manuscript Collection, University of Missouri, St. Louis.

At SVSU we received support from many individuals. Thomas Zantow, the head of Access Services cheerfully helped us get books and documents not available at our institution. Our student assistants, Amanda Jenkins and

Michelle Bell, tracked down many sources for us. Jennifer Paradise and Deb Roberts provided technical training and services when we needed it.

We also benefited from panel members at national conferences who read drafts of chapters. In that regard we wish to thank the panelists at the American Association of the History of Medicine Conference in 2008. Jason Scott Smith and panel members at the 2008 Policy History Conference made helpful comments on a paper about New York City's New Deal zoos. Robert Sugarman and other members of the circus section offered support and encouragement for a paper we delivered at the 2009 American Popular Culture/American Culture Association conference.

Ken Kawata and Robert Leighninger read early drafts of the book, catching many errors and making useful suggestions. Ken Kawata also shared resources from his own collection, connected us with members of the zoo community, and cheered us on in our effort to make sense of this era of zoo history. Thanks to Ken's introduction, Ray Pawley fact-checked an early draft of the snake chapter and gave us a mini-tutorial in herpetology. In those places where we have failed to follow our readers' advice, we hope that they will forgive us.

Finally we would like to thank our children, Miles and Maia, for enduring so many conversations about zoos with great patience.

Preface

In 1925 when he was 39, William Mann accepted the directorship of the National Zoological Park in Washington, D.C. At that time the zoo possessed "1,619 animals, representing 517 species." Mann's Ph.D. from Harvard, his former employment with the federal government as an entomologist, and his animal collecting trips around the world suggested that he would do a good job of managing and expanding this collection.[1] Although the National Zoo was among the nation's most prestigious zoos, it was not particularly well-funded, and from the start Mann's ambitious plans were curtailed by the reality that he would have to run the zoo on a very small budget, one that would not allow him to purchase, say, a gorilla. By the late 1920s, public zoos across the nation faced reduced operational budgets, and many desperately needed more than just maintenance. They required new exhibits and buildings. The National Zoo was no different, and it struggled to get adequate funding from the District of Columbia, which had other, perhaps more pressing, services to support. Mann did secure a special Congressional appropriation for the construction of a new reptile house before the stock market crash of 1929, but his zoo needed additional capital improvements. Mann dreamed of building new, barless animal enclosures, but after the District had appropriated funds for the police, fire department, and schools, he found there was never "enough money for the zoo."[2] Ironically, for Mann and nearly every other zoo in the country, the Great Depression proved to be a savior.

Thanks to a series of federal work relief programs initiated by President Franklin Delano Roosevelt — the Federal Emergency Relief Administration, the Civil Works Administration, the Works Progress Administration, the Public Works Administration — the National Zoo in Washington, D.C., actually expanded during the 1930s. With funding from these agencies, Mann was able to add an elephant house, a new wing to the bird house, a small mammals house, a zoo restaurant, and assorted shops and garages. Relief artists ensured that these structures were not just functional, but also beautifully adorned. Mann called each of the new buildings "the finest of its kind (I think) anywhere." Construction at the zoo was essentially ongoing until 1941. Lucile Mann, his wife, and an unpaid employee of the zoo, recalled that "the only time Bill really got a lot of money was during the [Great] Depression." The money kept Mann so busy overseeing construction and planning the next phases, that not until 1937 did he find the time to venture out on a collecting expedition to populate the new spaces.[3]

William Mann's experience at the National Zoo paralleled that of other zoo directors around the country in many, although not all, of its details. Like other zoo directors, Mann joined the zoo world during the roaring twenties, a decade in which scores of new zoos were established across the United States. There was a popular enthusiasm for zoos, and on paper it appeared that postwar America was becoming zoo rich. In reality, local governments found it difficult to support more than modest menageries, small collections of animals housed in temporary or inadequate cages. Zoo directors around the nation hoped and pleaded for money to expand their facilities and improve the welfare of the animals in their care, often without success. For them, as for Mann, the Great Depression and the arrival of New Deal work relief programs provided the opportunity to go on a building spree and, at a deeper level, to expand the public mission of zoos.

The fact of New Deal construction has been mentioned in passing by various students of America's zoos, and some single-zoo histories have included chapters on the era, but there has been no scholarly work devoted solely to studying how the American zoo community was transformed during the Great Depression.[4] Moreover, the stories of zoo professionals from this era have been obscured by a literature that tends to "dismiss [them] and their achievements as irrelevant" to the modern zoo's preoccupation with conservation and animal welfare.[5] This book seeks to fill that gap by considering a range of questions about the place of zoos in American culture and politics during the 1920s and 1930s. How, exactly, did zoo proponents convince local politicians and federal bureaucrats to direct relief money toward zoos? How did they defend such appropriations from critics? What did they do about animals they could no longer afford to feed? How did they get new animals when others died? How did the public view and use zoos? What kind of people ran zoos, and how was their profession affected by the Depression? What happened to zoos and their animals when America entered World War II and the New Deal programs came to an end? Did the New Deal leave a lasting mark on America's zoo landscape, or was its influence temporary? In considering these questions, we explore the nexus of culture, history, and politics of zoos across the country during this period.

We begin with the story of physical construction undertaken during the Depression. The vast majority of American zoos were public institutions funded by local government. Local parks departments typically ran city zoos, devoting a portion of the parks budget to their upkeep. Despite the ambitious dreams of their boosters, most city zoos were little more than collections of animals set in the middle of wide-open parks. Only a few elite institutions such as the Bronx Zoo, the Philadelphia Zoo, and the St. Louis Zoo had large, master-planned grounds with substantial permanent exhibits, but even some of these were in need of repair or under construction. Other major cities such as San Francisco and New Orleans had only modest zoos. After the election of FDR, the availability of relief labor through New Deal programs transformed the zoo landscape. Cash-starved communities turned to the federal government to either build new zoos, fix failing infrastructure, or complete projects started in the 1920s. Nearly every zoo in the United States took advantage of New Deal relief labor programs, and at least a score of entirely new zoos were constructed. It is no exaggeration to say that the 1930s were a golden age of zoo construction.

The physical transformation came during a period in which the zoo profession was itself changing. Chapter Two examines the professionalization of zoo directors and employees. Under the leadership of a few directors, a national zoo association was formed in 1924, and its members — many young and college educated — worked to promote a scientific, educational, and conservation mission for all zoos. This mission was already deeply established at a few elite institutions such as the Bronx Zoo, but the New Deal helped spread this agenda across the nation as new zoo directors made creative use of New Deal resources to build research laboratories, hire educators, and promote conservation messages. Guided by their national association, zoo directors worked to get local civil service boards to classify zoo employees differently from other city workers, giving them a professional status to match their specialized skills. To fill their new zoos, some directors tapped into a network of wealthy Americans who owned large yachts and could take curators on free collecting trips around the world and write off the trip as a scientific expedition for tax purposes. Even in the absence of such elite connections, zoos relied on an ever-growing trade network made possible by the development of a national community. The increasingly elite zoo professionals, however, did not lose touch with the public.

It is easy to imagine that zoos would receive very little public support during an economic depression. Why, in the context of massive unemployment and pressing social needs, should taxpayer money be spent on zoos and their animals? Indeed, vague reports that zoo animals were killed and eaten by starving citizens have made their way into some accounts of the Depression's horrible effects.[6] In Chapter Three we unpack the counterintuitive fact that public and political support for zoos was actually quite high. True, angry citizens did write to local newspapers questioning the wisdom of spending money on animals rather than on destitute humans. Others wanted to know why federal or local money was appropriated for zoos rather than for other cultural institutions such as libraries or art museums. But, consistently, politicians backed zoo projects on the grounds that zoos provided wholesome recreational and educational opportunities for average Americans dispirited by the Depression. Although zoos might have been run by an increasingly elite group of professionals, their appeal crossed all class lines. Politicians, eager to cater to the "everyman," found it easy to champion zoos. Civic pride and self-interest played a role too, as city boosters argued that a good zoo would attract tourists and their dollars. And, since federal relief projects employed local citizens, zoo improvements put people to work. Some proponents even used the fear of idle unemployed to argue that zoos could prevent crime by getting people off the streets. Only the cranks and Scrooges, it seemed, wanted to deny the people their right to a bit of free relaxation and entertainment.

Crowds flocked to zoos in record numbers, and among their most popular exhibits were the Reptile Houses. These "houses of horror" — as newspapers like to sensationalize them — appeared in many city zoos, thanks to New Deal funding, and they perfectly satisfied the general public as well as the growing ranks of herpetologists on zoo staffs. A fair number of zoo directors and curators, including the famous Marlin Perkins, had collected snakes as youngsters and specialized in herpetological studies. Their training, along with the fact that snakes were relatively inexpensive to keep, led them to develop impressive reptile collections. A number of reptile curators became national authorities on snakes, writing for both specialists and the general public, and providing a steady flow of snake stories to newspapers. In addition to publishing scientific information about reptiles, these curators worked to protect snakes from wanton human destruction and people from snakebites. Prior to the 1920s, there was no effective treatment for venomous snakebites in America, so herpetologists engaged in a public health campaign to educate Americans about how to identify and avoid venomous snakes. They also joined forces with a Brazilian scientist, university professors, a pharmaceutical company, and military personnel to develop a national antivenin network. Reptile houses and their curators played key roles in this network both by supplying raw venom and also by distributing antivenin to snake bite victims.

In a period of significant economic distress for humans, the welfare of captive exotic animals was defended and improved by zoo directors, as we explore in Chapter Five. New and bigger exhibits funded by the New Deal enabled animals to live in safer and healthier spaces. "Animal welfare," however, meant something different in the 1930s than it does to today's animal rights activists. At a fundamental level, an adequate diet was a measure of welfare, so directors protected their animals from the ravages of budget cuts by planting gardens, acquiring surplus produce from local farmers and markets, and running on-site slaughter houses to process horses and other animals donated by or bought from local humane societies. Directors also spent a lot of time defending zoo animals from being poisoned, taunted, and even shot by ignorant or malicious visitors. They joined forces with humane societies to lobby local politicians for physical improvements. In short, the New Deal dramatically improved the lives of exotic animals in zoos around the United States.

In our final chapter, we explore the end of the New Deal and its legacy for America's zoos. The United States' entry in World War II curtailed FDR's work relief programs, temporarily ended collecting expeditions abroad, and imposed a new set of hardships. The Manns first felt the gathering war's effects on their 1940 collecting trip to Liberia, which was complicated by Germany's aggression in Europe; for a while, they were not sure they would be able to return to the United States. When they returned, New Deal building programs were shut down, and many animal keepers joined the military. At the National Zoo, a new veterinarian was drafted, and even Mann spent three months in the South Pacific as a "technical observer" for the quartermaster corps.[7] Attendance dropped off too as gasoline rationing limited Americans' recreational driving.[8] These and similar effects were felt at zoos across the nation. Although the Depression ended and the war was soon won, the 1940s were not easy on zoos. Even new zoos wore out quickly, and few local governments could afford to maintain them properly; thus, by the war's end, many exhibits were a decade old and beginning to need repairs. Moreover, changing standards of animal welfare meant that some of the most "modern" exhibits of the 1930s were judged to be wholly inadequate only a few decades later. Zoo budgets stagnated or declined, and by the time of Mann's retirement in 1958, even the National Zoo was beginning to explore non-governmental sources of support. By the 1970s, a fair number of zoos were on the brink of closure. This time, however, no federal program emerged to save them. Instead, a new era of private-public partnerships generated the capital needed to restore and expand the New Deal zoos. Today, the legacy of the New Deal is still evident—in lovingly restored and remodeled buildings, in art, and in the leadership of zoo professionals whose careers were made possible by a massive government investment during a single decade so long ago.

We examine this period largely through archival material drawn from across the United States. We do not pretend to offer a comprehensive analysis of *every* zoo affected by the New Deal; this is not a quantitative study.[9] Moreover, we have given little space to some of the zoos (e.g., Toledo, Detroit, or Chicago's Brookfield) whose New Deal history has been covered in other books. Instead, we have tried to focus on zoos whose experiences speak well to the patterns of change felt across America during this time. Our zoos represent the geographic diversity of zoos affected by the New Deal, as well as the diversity of local political conditions. Our evidence comes from archival collections at the Smithsonian, the University of New Orleans, the New Orleans Public Library, the New York Public Library, the New York Park Parks Department Library, the New York City Municipal Archives, the Staten Island Zoo Archives, the Dallas Municipal Archives, the Western Historical Manuscript Collection at the University of St. Louis (Missouri), the Seattle Municipal Archives, the Woodland Park Zoo Archives, and other locations. We also use newspaper accounts, memoirs, contemporary scholarly works, and government documents. We hope that our book will contribute to conversations and further research about the role of zoos in our culture, as well as about the legacy of the New Deal.

ONE

Building the New Deal Zoo

The spread of parks, playgrounds, and parkways in New York City and the Metropolitan area since January 19, 1934, is one of the silver linings of the depression.

— Robert Moses, 11 Oct. 1937[1]

America had a relatively small number of public zoos in 1932, the year that Franklin Delano Roosevelt was elected president, and most were modest. In that year, the American Association of Zoological Parks and Aquariums (AAZPA) conducted its first survey of the nation's zoos. It managed to locate 58 zoos that provided data on their collections. Another eight zoos provided no data, suggesting that they were in their infancy.[2] In the same year, the American Association of Museums published a guide to the nation's zoos, aquariums, and botanical gardens. This guide listed 88 public zoos and 10 privately owned zoos, representing 35 states.[3] Some states had more zoos than others. California, Michigan, Minnesota, New York, Ohio, Pennsylvania, and Texas each had five or more.

These zoos could be divided into at least three categories. At the top were the elite zoos whose annual operating budgets were over $100,000. These seven zoos — San Diego, National (Washington, D.C.), Chicago, Lincoln Park (IL), Detroit, St. Louis, Bronx — had national and even international reputations. Physically, these were also unusually large zoos, occupying over 100 acres each, with the exceptions of Lincoln Park (15) and St. Louis (77); the typical zoo was less than 50 acres. And, zoos in this class had large and varied animal collections, as well as numerous permanent buildings. One large step down were 25 zoos with budgets between $20,000 and $100,000. A handful of these zoos compared favorably with the top zoos in terms of the quality of their collections and professional management, but most were mid-sized zoos owned and operated by mid-sized cities. A dozen of these got by on annual budgets that were one-tenth the size of those enjoyed by the wealthiest zoos. The majority (50) of the zoos identified in 1932 fell into a final category that we could label small, indifferent, or struggling. These zoos had budgets of $20,000 or less, with 16 getting by on less than $5,000 a year, and most had very small staffs, with some employing only a single keeper or two.[4]

What these figures do not reveal is the precarious situation facing many of these zoos, even the wealthier ones. The 1920s had been a great era for establishing new zoos, but the physical development of these zoos had been curtailed by the Depression. Older, bigger, and more established zoos found themselves struggling as well.

Relief labor provided by a number of New Deal programs did not just build new zoos; it also sustained existing ones during a period of prolonged national fiscal crisis. For some, the Depression ended up being their salvation. Without the massive infusion of federal funding, it is likely that some zoos would not have survived the Depression. Even without the Depression, however, it is possible that some zoos would have failed.

The 1920s — Unrealized Dreams Dashed by the Depression

Many municipal governments across the nation supported modest public menageries prior to the 1930s. Usually located within an existing city park, these early zoos typically lacked a perimeter fence, giving the public unimpeded access, often by automobile, to haphazard collections of animals. Sometimes started with cast-off circus animals, these menageries grew at an uneven pace, guided with varying levels of attention by a park department director, by a park commission or board, or by a group of citizens organized into a zoological society. Public and private funding for such zoos ebbed and flowed, with a new building built now and then, and new animals added on occasion. The uncertain status of these animal collections is obvious in the official documents, which variously refer to them as "zoos" and "menageries." In the parlance of the day, a "menagerie" signified a collection of caged animals with no pretense of education or science. When managed by city parks departments, such menageries used regular park employees for most or even all of the maintenance and feeding duties. Thus, zoological societies and park boards spoke often of their desire to transform their "menageries" into proper "zoos" or "zoological parks." Such dreams, however, were initially dashed by the nation's economic crisis.

The Dallas Zoo (known prior to the 1950s as the Marsalis Park Zoo) had a typical pre–Depression history. The city of Dallas purchased two deer and a mountain lion to exhibit in City Park in 1888, adding some bears a few years later. Fears about liability and upkeep costs led the city council to propose selling the bears, but a public outcry saved the zoo, which struggled along, moving from one location to another, getting by on annual city appropriations of as little as $25 a year, and exhibiting an unexceptional assortment of deer, raccoons, squirrels, and other ordinary animals. The public, however, visited in large numbers, and by 1912 a group of influential citizens pressed the city to devote more resources to the zoo, which it did. Children donated pennies, nickels, and dimes to purchase two elephants, one in 1915 and one in 1922. Also in 1922 a Dallas Zoological Society was formed with the goal of building "a first class Zoological Garden." The society raised $12,000 to fund the acquisition of 60 different species from famous animal dealer Frank Buck. By the mid–1920s, the zoo had improved dramatically in terms of the animal collection, but the exhibit spaces were still quite modest. The opportunity to improve the physical space was lost with the economic downturn of the late 1920s, and by 1932 the Park Board felt forced to sell off a number of animals.[5] And, because the animals reproduced in captivity, and citizens were constantly donating new ones, the zoo's animal population was growing.

Even a large, prosperous city like New York City had similarly poor quality public zoos, although its Bronx borough was home to the New York Zoological Park (better known as the Bronx Zoo). The Bronx Zoo received some financial support from the city, but it was managed by the private New York Zoological Society (NYZS)—a group of elite men with philanthropic impulses and conservationist convictions.[6] The zoo was—like the city itself—large in scale and ambition. Covering 264 acres, it was the world's largest zoo when it opened in 1899, and retained that title among American zoos, claiming 29 permanent buildings and nearly ten miles of roads and walks by 1939. Its animal collection was unmatched—both in size (over 1,000 species) and quality. Its scientific and conservation orientations (the society published both a popular magazine and a quarterly scientific journal) put it in the company of a handful of elite institutions. And the public loved it. Somewhere between two and three million people walked through its gates every year, paying not a dime unless they visited on Monday or Thursday, when admission cost adults 25 cents and children under twelve 15 cents. For New Yorkers with more aquatic inclinations, the society also ran the New York Aquarium, located within a remodeled historic fort in Battery Park, at Manhattan's southern tip.[7] As had been done with the Bronx Zoo and other museums, the city chartered the aquarium (in 1896) and then transferred management responsibilities to the New York Zoological Society, which received an annual operating budget from the city but raised its own funds for the exhibits. In 1936, the New York Aquarium was claimed to be the second largest in the world in terms of physical space, and the largest in terms of number of species and specimens. And, like the Bronx Zoo, the institution pursued a vigorous scientific agenda, maintaining a research laboratory, raising game fish for New York State waterways, and publishing a book series dedicated to aquatic life. The aquarium's two-and-a-half million annual visitors, however, were likely attracted by the spectacular setting, free admission, and entertaining collection.[8]

New York City's two other zoos bore little resemblance to the Zoological Society's world-class institutions. Manhattan's Central Park Menagerie could brag that it was the oldest American public zoo, but it exhibited a motley collection of cast-off circus animals, unmanageable pets, and donated creatures both exotic and common. Operated by the Parks Department's Manhattan division and housed in mismatched and poorly maintained structures, the menagerie at the turn of the twentieth century was described as a "ghastly place" best visited only in the winter when the urine stench was somewhat less powerful and the rat infestation somewhat less visible. Even as the menagerie's physical decline continued, the animal collection grew steadily and attracted large crowds of New Yorkers — as many as 20,000 on weekends — eager to get close enough to lions, bears, tigers, sea lions, and monkeys to feed them, or perhaps to take a pony or camel ride. The Menagerie benefited from minor repairs in the 1920s, but it was still a substandard, unsanitary zoo.[9] One borough to the east, Brooklyn's residents were able to visit a small, informal menagerie established in Prospect Park in the late 1800s. Pens and open cages held sheep, dogs, deer, rabbits, geese, and similar semi-domesticated animals until the 1910s when the Brooklyn Zoological Society was founded to give the park a more exciting and permanent zoo. By 1920, visitors could find lions, monkeys, and bears, some inside a modest new brick building.[10] The zoo was looking up, but it was a long, long way from the standard set by the Bronx Zoo. And, with the belt-tightening required by the Depression, the future prospects for both public zoos looked grim.

The Cincinnati Zoo was one of the nation's largest, most famous, and oldest zoos (established in 1874), but it too was in poor shape when the Depression hit. Nearly all of its major buildings were close to 60 years old, and its funding system was similarly antiquated. Unlike most American zoos, Cincinnati's was both owned and managed by a zoological society, not the city, and it had always charged an admission fee to support its operations. This admission fee — 25 cents for adults and 10 cents for children — had not been raised since the zoo opened, and as expenses mounted during the 1920s, revenues did not keep pace, preventing any new construction during the supposedly flush 1920s. Once the Depression hit, even 25 cents was too much for many, and annual attendance dropped along with annual revenues.[11] Facing budget deficits, the zoo agreed to a deal in 1932 whereby the city of Cincinnati gained title to the zoo itself, while the Zoological Society retained management power.

The city of San Antonio solved its budget woes with a similar shift of fiscal responsibility. In San Antonio's case an active zoological society had been frustrated throughout the 1920s by the city's annual poverty plea. Plans to develop barless exhibits and add animals ran up against the city council's refusal to appropriate funds for such purposes. To break this impasse, the zoological society negotiated an informal understanding in 1929 whereby it would raise money only for animals and physical improvements, while the city would be responsible for basic operating expenses. Formalized a few years later, the agreement gave the society 50 percent of gate receipts and all revenue from concessions. A real zoo began to take shape, but the Depression limited the society's fundraising abilities.[12]

The New Orleans Zoological Society reported that 1928 had been "a most successful one in the Zoo," with the construction of a new sea lion pool, a reptile house, and a new lion cage. The success masked underlying problems, however. The bird cage needed repairs. The animal collection was growing, but the city did not come through with an addition to the annual budget, meaning that "all available funds are consumed to take care of the animals and birds which we have now." By the end of the fiscal year, the zoo had gone $5,000 over budget. Ten years later, the structures that had seemed more or less fine were recalled as "old unsightly wooden cages."[13]

Many municipal zoos entered the Depression in a precarious position because they were so new, usually under a decade old. Established by civic-minded citizens, they were in their physical infancy, often lacking the kind of permanent infrastructure that would have made them real in the minds of local politicians. It was much easier to justify paring back zoo expenses when there was very little actual zoo to see. Duluth, Minnesota, was in this situation. Its Lake Superior Zoo began in 1923 with a single pet deer housed on city property adjoining the lake. Two years later the zoo's collection had expanded to include some buffalo, foxes, wolves, coyotes, bears, and pigeons. A single building was constructed in 1927 to house the collection, with cages in the basement for big cats, some of which were purchased in the same year. The zoo grew a bit, and then the Depression forced serious discussions about closing it.[14]

In Tulsa, Oklahoma, similar bad timing affected the municipal zoo, but a highly dedicated staff kept the zoo running. Opened in 1928 at the tail end of an oil boom, the zoo went on a "Depression diet" almost immediately, with zoo staff scrounging spoiled vegetables, picking up road-kill, and harvesting park grass and trees to feed their 35 animals. As Tulsa's oil industry faltered, the city's population stagnated and the unemployed went on their own involuntary diets. Although one zoo building had been completed in 1929 and a monkey island by 1931, the city budgeted barely $200 a month for feed; the zoo survived the first years of the Depression thanks to an unusually professional staff. Park Superintendent Will O. Doolittle had been mentored by the National Zoo's first director, William T. Hornaday, and was a leader of the AAZPA, which was then affiliated with the American Institute of Park Executives (AIPE). Doolittle wrote zoo-related articles for the AIPE's journal, *Parks and Recreation*, and in 1932 he compiled the first directory of American zoos and aquariums. The zoo's director, Hugh Davis, was a conservation advocate and an active AAZPA member, and other zoo staff apparently shared the two men's commitment to making the zoo into a true educational center.[15] Idealism kept the zoo and its animals alive.

Pre–Depression zoos struggled for several reasons. The most obvious were financial. Only a handful of American zoos charged admission, and those tended to be the oldest, largest, and privately-managed ones, which had national and even international reputations. The Philadelphia, Cincinnati, Bronx, and San Diego zoos could justify a small gate fee because their large collections and park-like settings were worth the cost of admission. Even a modest admission fee raised a substantial amount of money because these zoos had hundreds of thousands, even millions, of visitors each year. Zoo concessions facilities were owned by the zoological society, which kept the profits, unlike municipal zoos, which tended to receive only an annual fee or profit percentage from concessions contractors. In addition to gate and concession revenues, these zoos enjoyed some fiscal support from local government, even though each was owned and managed by a private zoological society. Finally, wealthy zoological society members often donated large sums for special purposes such as the construction of new exhibits. With this steady income from various sources, the zoos were better situated than their poorer municipal cousins to hire highly qualified staff, make needed repairs, and build new exhibits.

By contrast, all but one municipal zoo in 1932 was free, meaning that they were almost entirely dependent upon annual appropriations from city government. These appropriations tended to be small, enough to pay a few employees and feed the animals, but not much more. Private benefactors and citizen subscriptions paid for the occasional new animal or exhibit, but these were not reliable income sources. The problem was not that politicians did not like zoos. Rather, local governments had limited revenue sources, and in pre–Depression America small government was a virtue. Low taxes and expenditures were the norm, and there was often little political will to change that situation, no matter how much citizens loved their local zoo. Smaller zoos with active zoological societies were somewhat better positioned, but they too had difficulty raising the substantial amounts of capital required to turn their visions into reality.

Finally, America's smaller zoos were frequently run by amateurs who lacked the desire and knowledge to lobby effectively on their institutions' behalf. A majority of municipal zoos were managed by the local park department, with the park supervisor also serving as the zoo director. Regular park employees were assigned to take care of the animals. Such staffing arrangements could mean that those people who actually ran the zoo gave it partial or indifferent attention. No one had a clear stake in the zoo.

So, even before the Depression hit, all but a few of America's zoos were on a razor's edge. A budget cut, a drop in donations, or a combination of both would mean that animals could not be fed and exhibits could not be maintained properly. Large crowds of eager visitors masked inadequate budgets and infrastructure. Civic-minded Americans had established zoos at a dizzying pace during the first two decades of the twentieth century, but they had done so without serious consideration of the costs associated with maintaining a first-rate zoo that housed its animals safely and humanely. The concept of a "master plan," which implied large capital outlays and long-term support mechanisms, was rarely heard amid the enthusiastic cheers to build a "first-class" zoo. For most communities, a "first-class" zoo was just so much rhetoric.

The onset of the Depression exacerbated underlying financial and management problems. Park and zoo budgets were slashed. In 1935 Milwaukee's Washington Park Zoo had $4,000 cut from its $28,000 operational budget. Employees' wages were reduced or not paid at all. Donations fell off. Directors went begging. Park boards began worrying about how they would feed the animals. The Audubon Zoo feared it would have to cull three of its sea lions to ensure that all had enough fish to eat.[16] Edmund Heller worried that his monkeys would suffer without bananas.[17] Projects in-progress came to a halt. In 1937 the Brookfield Zoo stopped work on its monkey island due to lower-than-expected tax revenues.[18] As budgets and private donations fell, the future of America's zoos was uncertain. In St. Louis, opponents of a measure to reduce a zoo tax feared that such "Measures of economy in a troubled year can expend their force in other ways than in attack on two of our show places. Once reduced each tax might not be restored for years, or might never be restored."[19]

Facing budget shortfalls and unable to collect gate revenues, zoos adopted a number of cost-saving measures. Animals went on what Ed Bean, director of the Brookfield Zoo, called a "Depression diet." If they did not already have

Exterior of restaurant, Audubon Zoo. The construction of permanent, sit-down restaurants in New Deal zoos provided "upper class" leisure for the public (Louisiana Division/City Archives, New Orleans Public Library).

them, parks designated space for raising fresh vegetables to feed the animals. Detroit, Chicago, Seattle, and Tulsa all had these zoo farms. Less-than-fresh fruits and vegetables destined for the landfill were collected from grocers and markets in Tulsa and other cities. Seattle's Woodland Park Zoo employees, for example, made regularly scheduled trips in their 1½ ton Dodge truck, collecting tens of thousands of pounds of food each year. Grass cut in city parks was used as a hay substitute.[20] Dead and condemned animals from the local animal shelter were butchered for the big cats and other carnivores. The health effects of this cost-cutting could be bad, according to some critics. A Central Park zookeeper claimed that his cheetah, "Bongo," developed distemper and sore eyes because the city had cut venison from his diet to save money.[21] It seems more reasonable, however, to argue that these various thrift measures were less efficient than they seemed. Zoo and park employees spent a lot of time collecting and preparing food that in flusher economic times would have been delivered directly to the zoo. The unpredictable food supplies surely affected keepers' ability to regulate their animals' diets. Later generations of zookeepers understood that donated foods such as stale bread would be eaten by animals but lacked nutritional value.[22]

How the New Deal Built Zoos: The Programs and Their Impact

The Depression threw America's zoos into dire straits and then the New Deal rescued them. Paradoxically, the New Deal built zoos both rapidly *and* over a long period of time. A succession of relief programs — the Public Works Administration (PWA), Federal Emergency Relief Administration (FERA), Civilian Conservation Corps (CCC), Civil Works Administration (CWA), Works Progress Administration (WPA), National Youth Administration (NYA), and Federal Art Project (FAP) — were utilized by zoos, but each program operated with incredible urgency since its authorization and appropriations were reviewed by Congress each fiscal year. Want a new reptile house? Get the project authorized and work quickly, because when the relief labor disappears, the construction will halt. Many of the earliest New Deal zoo projects were begun by CWA laborers, continued under the FERA, and

Temporary "USA WORK" signs reminded citizens of the employment benefits provided by New Deal zoo construction projects. Permanent bronze plaques were often installed as well, and many can still be seen today (Louisiana Division/City Archives, New Orleans Public Library).

completed by the WPA. Once WPA construction crews finished their work, artists from the FAP might add sculpture, ornamentation, murals, and signage. Young men and women from the NYA would later paint buildings, build benches, complete landscaping, and assist zoo staff in a variety of roles. A supplemental WPA project could be used to improve sewage or drainage infrastructure, create permanent walkways or parking lots, or even remodel structures erected just a few years earlier. Thus, although relief labor built new zoo buildings with amazing speed, workers from a range of New Deal programs lingered for years after the official opening date, sustaining the continuous improvement and maintenance of America's zoos until the nation's entrance into World War II. Properly understood, the New Deal supported America's zoos throughout the Great Depression, not just with one-shot building projects.

During President Roosevelt's first 100 days, the PWA received a $3.3 billion appropriation under Title II of the National Industrial Recovery Act. Directed by Harold Ickes, however, the PWA moved slowly to develop projects and utilized some private contractors, rather than just the unemployed. Eventually, the PWA was responsible for numerous and notable infrastructure projects, including some for zoos and aquariums in St. Louis, Cincinnati, Dallas, and elsewhere.[23] In the short term, however, the PWA did too little to address the nation's unemployment crisis, so it was supplemented with other programs designed to put people to work quickly.[24]

To "aid the states in meeting the cost of furnishing relief and work relief, and in relieving the hardship and suffering caused by unemployment," the FERA was created on May 12, 1933. By June an administrative structure for distributing funds was in place, and the FERA administrator, Harry Hopkins, met with governors and state emergency relief administrators (whose positions had been necessitated by the earlier Reconstruction Finance Corporation's requirement that each state create a relief agency to distribute federal loans and grants) to explain how the new program would work. The FERA represented a major shift of relief work away from private organizations and toward local and state government. Under the RFC, some federal funding had been directed by the states towards the private charitable and relief organizations that had traditionally handled such work; under the FERA "it was stipulated that public funds should be administered only by public officials administratively responsible to regular units of government." The FERA Administrator gave grants to the states, which then combined the federal money with state and local funds either to provide direct relief or to finance work-relief projects. The latter kinds of projects were favored because they reflected the FERA philosophy that "public works are the most obvious governmental device for relieving private unemployment." Public works projects had the advantages not just of providing employment but also of leading to the creation of tangible modifications of the landscape — buildings, roads, playgrounds, etc. — that were of "high social value and benefit to the community."[25] Because the FERA was designed to maximize employment, it also limited the number of hours that an individual could work, restricting each to just the number required to supply his or her "minimum budgetary requirements"; as a result, local project administrators were forced to employ workers in shifts, a requirement that created some inefficiencies.[26]

Harry Hopkins, the FERA administrator, hoped for a program that would be a purer public works vehicle. He convinced FDR to create the Civil Works Administration, which employed four million Americans during the winter of 1933/1934. The FERA resumed responsibility for unemployment relief when the Civil Works Administration was discontinued (April 1, 1934) and continued through July 1, 1935, when it was replaced by the Works Progress Administration.[27] The 1935 passage of the Social Security Act meant that direct relief for the unemployable would no longer be the responsibility of the FERA's successor, the WPA. Instead, the WPA would focus on creating jobs for the unemployed.[28] Many of these jobs were on projects that built local infrastructure, including zoos.

Other New Deal agencies made significant contributions to zoos. Crews of young Americans working for the Civilian Conservation Corps assisted zoo development. In Tulsa they built an $8,000 zoo refectory building.[29] In Mesker, Indiana, CCC crews essentially constructed the Mesker Park Zoo as part of a larger parks project. The National Youth Administration also contributed crucial labor to zoos. After WPA crews left a job site, many details such as trim painting and landscaping might remain unfinished. Two years after the new Audubon Zoo opened to the public, NYA painters began work on rails and other elements in "all" of the buildings.[30] At the Staten Island and National zoos, young men and women from the NYA provided technical skills such as preparing specimens, making reports, and creating models.[31]

The federal architects of the New Deal programs publicized their agencies' accomplishments in reports that argued the purpose and efficacy of the thousands of local projects. The reports stressed what Jacob Baker, the assistant administrator of the Federal Emergency Relief Administration, called "the essential facts."[32] These facts, collated from state reports, emphasized numbers: how many projects built, how many workers employed, how many citizens served, and so forth. The numbers were supplemented with photographs of completed or in-progress public works projects.

Zoos constituted a small percentage of total relief projects, but it appears that a majority of zoos benefited from New Deal programs, and a fair

Artists chisel a Noah's Ark bas relief on the Audubon Zoo's bird house. The FAP sculptor, John McCrady, won a Guggenheim in 1939 (Louisiana Division/City Archives, New Orleans Public Library).

number enjoyed dramatic reconstruction. In 1935 the FERA reported that "work on the construction and improvement of 68 zoos was done in 18 states."[33] Only 88 public zoos existed in 1932, so over 80 percent of America's zoos had been affected by FERA alone. At least eight new zoos opened between 1933 and 1940, and it appears that all of these were built by the New Deal. Several zoos with pre–1933 establishment dates were actually constructed or completely rebuilt by the New Deal: the Staten Island Zoo, San Francisco Zoo, Buffalo Zoo, Prospect Park Zoo, Central Park Zoo, and Toledo Zoo.[34] So, it is probably not an exaggeration to say that every public zoo in America benefited from federal relief labor during this period.

Although New Deal programs sought to spread their benefits through every state and county in the nation, some politics were involved in the actual allocation of construction funds. When Roger Conant arrived at the Philadelphia Zoo in 1935, he was shocked to discover that it had not been significantly improved by any federal programs, a fact that he attributed to the local "Republican stronghold" that had "been reluctant to take FDR's tainted money."[35] On the other hand, governors and local political leaders sympathetic to Roosevelt's relief programs were more likely to develop projects that could take advantage of New Deal labor.

For example, New York City's Mayor Fiorello La Guardia, elected in 1933, was an enthusiastic supporter of any federal programs that promised to help him improve the lives of the city's residents. His newly appointed Park Commissioner, Robert Moses, may have harbored some doubts about the efficiency of the federal relief programs, but he also recognized that they would help him embark on a massive construction spree that would transform not only the city's parks but also its transportation arteries. He had no special love of zoos *per se*, but they proved to be useful publicity tools in his broader project to transform the city's physical landscape. Mayor La Guardia came to power at just the right time to take advantage of these federal programs. His vision for the city included improving its parks "to give people something which will not only be of recreational value but also be esthetically stimulating." Unfortunately, he took office without the "chisel or brush" to make real this dream.[36] La Guardia found his chisel and brush when he appointed Robert Moses Park Commissioner on January 19, 1934, the same day that the governor signed legislation that combined New York City's five borough park departments into one, giving Moses significant power to shape the city's public landscape.[37] Moses was a Republican who had served in Governor Franklin Roosevelt's administration, and La Guardia imagined that he would have the political connections to tap into New Deal funds. La Guardia did not realize that Moses and FDR had become enemies, although it turned out that Moses possessed other political skills that were more useful in the long run.[38] For his part, La Guardia was on friendly terms with FDR and his New Deal administrators and became quite adept at keeping the federal money flowing; in a 1937 speech at Brooklyn College, FDR quipped, "Every time La Guardia comes to Washington I tremble because it means he wants something, and he almost always gets it." La Guardia felt tremendous pressure to bring home federal assistance to a city where one out of five residents was on relief in 1936.[39]

Another factor that affected which cities made significant zoo improvements was advance planning. Requests for relief work funds were evaluated by engineers who placed the highest priority on those requests that were accompanied by detailed construction plans. Less than a week after taking office, Moses faced down the CWA over its policies of hiring only unemployed men and of keeping design costs to a minimum. Moses wanted a large staff (600) of architects and engineers, and when he threatened to quit if the CWA would not bend its rules, La Guardia intervened and Moses got his way. Robert Moses' design team ensured that Mayor LaGuardia had a steady supply of blueprints to present to federal administrators (the team designed the entire Central Park Zoo in just 16 days), practically guaranteeing that New York city would get the funding it requested, since the federal administrators had a strong interest in supporting projects that would likely be completed.[40] Although Moses clashed frequently with FDR and federal relief administrators, he was an indispensable New Deal ally because he produced enormous numbers of public works projects in record speed and he was a vocal advocate of the WPA's conception of the role government could play in public recreation.

Cities whose zoos or park departments had made plans for major expansion prior to FDR's inauguration were thus well poised when federal funds became available. In the Toledo Zoo's case, Roger Conant, a young and ambitious reptile curator with big dreams but no reptile house, worked through a Zoological Society board member to get a young, struggling architect, Paul Robinette, to donate some of his unused time to designing a proper reptile house for the zoo. When the regional CWA administrator came looking for projects in 1933, the zoo was perfectly positioned to launch a construction project. Quickly, Robinette drew up additional plans, and by the time the New Deal programs ended, they had funded the construction of six new buildings, completely transforming the Toledo Zoo from a typical municipal menagerie into a beautiful and substantial zoological institution.[41] Dallas got off to a somewhat slower start, but at the end of 1934, the Dallas Park Board had landscape architect Wynne B. Woodruff draw up a report of suggested improvements for the city's park system, complete with drawings. Within the next three years, New Deal programs allowed the Dallas Zoo to realize parts of Woodruff's plan.[42]

Because the CWA and WPA programs did not always pay for building materials, cities and zoos had to find creative ways to stretch their resources. The Toledo Zoo's WPA buildings were constructed largely with materials salvaged elsewhere in the city. In Buffalo, the new zoo was made from native limestone, some of which was quarried on-site when laborers dug pits for new exhibits. Local stone was also used in Pueblo (Colorado), San Antonio, and Washington Park (Indiana).

Most New Deal programs required some kind of contribution in materials or cash from a local sponsor. These contributions could come from local or state government, or even from private benefactors. The Cincinnati Zoo pursued the latter option. Owned and managed by a private zoological society prior to the Depression, the zoo's major benefactor had died in 1927, leaving the zoo nearly entirely dependent on its gate receipts, which were insufficient. So, in 1932 ownership of the zoo was transferred to the city's Park Board, and a new Zoological Society of Cincinnati assumed responsibility for funding and running the zoo's daily operations. As part of the agreement, another private benefactor paid the zoo's 1931 deficit, while the city paid the society $325,000 for the property itself. The newly organized society immediately set about tapping federal funds. By asking for funds from private donors, the society raised enough money to secure PWA, CWA, and WPA projects that allowed the zoo to expand during the Depression. Barless Hagenbeck–designed or inspired exhibits were added — African Veldt (1935), Lion and Tiger Grottos (1934), Bear Grottos (1937) — as were a beautiful $35,000 art deco Reptile House (1937), an animal hospital, and other exhibits and structures.[43] A number of the most prominent zoos in the nation, such as the Philadelphia, San Diego, and Bronx zoos, were private-public partnerships. They had been established by non-profit zoological societies that continued to manage and staff them, while the local government provided some support for operating expenses. All of these zoos used contributions from private donors to secure federal relief projects. Thus, not only municipal zoos benefited from New Deal funds.

The precise nature of federal involvement in zoo development was not always explained clearly in official reports. In 1935 the FERA report highlighted its zoo work by describing improvements made at Chicago's Brookfield Zoo. The report emphasized that the zoo was three times larger than its European counterparts, with 113 acres for exhibits and 46 acres for parking.[44] This statistical rhetoric somewhat exaggerated FERA's role in building the Brookfield Zoo, since only a few of the exhibits on the 113 acres were constructed under it or any of the successor federal programs. Still, federal relief workers were a huge presence at the zoo throughout the Depression. FERA laborers were supplanted by CWA and then WPA workers, as many as 125 at a time. Between 1934 and 1939, these relief workers built a parking lot, constructed nearly all the roads and walkways connecting the zoo's various exhibits, cleared land, excavated moats and grottos, constructed the metal lath understructures for the many artificial rock exhibits, and painted new buildings. Artists from the WPA's Federal Art Project carved directional signs, painted scores of murals in the reptile and elephant buildings, and designed zoo posters, maps, drinking fountains, decorative screens, table tops, and sculptures. Much of this work was done thanks to a single WPA authorization worth $254,000 (the Zoological Society, as the sponsor, contributed $43,000). Most of the actual exhibits — bear grotto, sea lion pool, etc. — were private contract projects.[45] Although the FERA exaggerated when it implied that the Brookfield Zoo had been built by federal funds, it is certainly true that the various New Deal programs literally made the zoo accessible to the public and certainly made it more beautiful. It is unlikely that the zoo would have been completed in the absence of these federal relief programs.

New Deal funds allowed cities to make labor-intensive improvements to their zoos that made them more accessible to visitors and more comfortable for their animals. The Dallas Zoo spent nearly $100,000 to build a system of concrete and native brown fieldstone bridges, paths, stairways, and animal cages and paddocks. By 1937 additional WPA work produced a stone administration building, a "Cat Row," and a monkey house, all still used today; within a decade, the city claimed its zoo was one of the nation's ten largest.[46] The city also acquired a modest aquarium in Fair Park, thanks to a Public Works Administration project.[47] San Antonio had WPA laborers at the zoo from 1935 to 1942, excavating barless enclosures for camels, monkeys, dik-diks, rhinos, hippos, and cranes; building a Peccary House, a bird of prey house, and a reptile house; and constructing numerous trails and walls with native limestone.[48]

Remodeling and enlarging an existing zoo could have a high price tag. For example, New York's Buffalo Zoological gardens benefited from a $2.8 million WPA "modernization" project.[49] New York City's three New Deal zoo projects were more modest, ranging from a half million to a million dollars apiece. But construction costs tended to balloon from their initial estimates. All three of New York City's zoos doubled in price during construction. The Audubon Zoo began as a $240,000 project, but when it opened the cost had apparently climbed to nearly a half million dollars. Yet, the "final" costs were slippery. Part of the price was borne by the federal government, part by the local sponsor, but news reports rarely distinguished between these. In fact, newspaper articles, presumably reporting figures given by city officials, often seemed to inflate the actual cost of projects. The Audubon Zoo's initial cost was reported in newspapers as $240,000, but the Park Commission's internal reports put the amount spent exclusively on the zoo at something more like $156,000. The higher figure would have been accurate for the expenditures on the *entire* park, but it was consistently linked to the zoo improvements alone. So, despite the Taxpayer's League and critics of "wasteful" New Deal programs, local politicians had an incentive to exaggerate the amount of money spent on their projects. In a nation where bigger has always been better, and in the context of massive unemployment, a high WPA price tag signified both that the city had added something really important and that it had employed many people in doing so.

To ensure that taxpayers recognized the major improvement to their communities made possible by New Deal programs, federal administrators encouraged the erection of "America Works — WPA" signs at construction sites, which was interpreted by some as a blatant political strategy to buy goodwill for FDR and the Democratically-controlled federal government. When a WPA bulletin suggested that permanent plaques and tablets should be used to commemorate WPA work in city parks, Moses wrote a sarcastic, two-page explanation of why he most certainly would do no such thing. After pointing out the obvious fact that many park projects had been completed with the support of multiple alphabet agencies, he objected to the plaques on aesthetic and political grounds. In addition to being ugly (the bulletin recommended that plaques be placed on "substantial concrete posts" erected at park entrances), such plaques represented an unlawful extension of federal authority over local decisions and an unseemly aggrandizement of the federal government. Moreover, his reference to a former mayor who had plastered the city with similar signs indicated that Moses saw the plaques as serving an illegitimate political purpose — to create support for the current administration.[50] In New Orleans a "taxpayer" complained to Mayor Robert Maestri about the ubiquitous WPA signs throughout the city, asking, "do we *have* to advertise to [visitors] that all this beauty was provided by the WPA? Why not give these folks the impression that we've always had a city beautiful?"[51] The truth was, however, that American cities and their zoos were being made more beautiful and useful by these New Deal projects.

The physical and aesthetic impact of the New Deal is best understood through the consideration of several zoos built entirely by federal relief laborers.

Three Zoos for New York City

When the stock market crashed in 1929, New York City's Wall Street marked the epicenter of the financial disaster. In pure numbers, one might even say that the spreading economic depression also hit New Yorkers hardest, since the metropolitan area, although only 229 square miles, was home to more people than the 14 smallest states combined, with three of its boroughs among the eight largest cities in America.[52] The sheer number of people concentrated in New York City meant that the Great Depression's effects there were dramatic. Its size also meant that it would take a lion's share of the federal government's assistance programs, and armies of relief workers labored on public works projects of all descriptions. Thanks to New Deal programs, several hundred thousand New Yorkers were hired to refurbish parks, build playgrounds, paint murals, construct swimming pools, and make numerous improvements to the public infrastructure. Not least among the city's federally funded projects were three new public zoos.

New York City in 1934 led the nation not just in terms of population, but also in its zoological reputation, with the Bronx borough being the home of the New York Zoological Park (better known as the Bronx Zoo), one of the world's top zoos.[53] New York City's two other zoos — Manhattan's Central Park and Brooklyn's Prospect Park zoos — were grossly inferior to the Bronx and no better than the scores of small-town zoos spread across the

United States. And, on Staten Island, a new zoo was in its infancy, a plan with a site but no permanent buildings. Thus, New York City at the onset of the Great Depression was home to one world-class zoo, two established but inferior menageries, and one planned zoo. Standing in the way of change were two major hurdles: a lack of money, and a borough-based park department system described by its critics as distinguished by inefficiency and corruption. The money problem was addressed soon after Franklin Delano Roosevelt assumed the presidency in 1933 and oversaw the development of federal agencies that would funnel large amounts of capital and labor to the states and municipalities for public works projects. Simultaneously, change at the leadership levels of New York City's politics created a new landscape in which new zoo construction became not only possible but also a priority. Two men — Mayor Fiorella LaGuardia and his Park Commissioner Robert Moses — pushed through sweeping changes in how park business was conducted. Within months of taking office in 1934, Mayor LaGuardia's administration had utilized New Deal resources to completely rebuild the Central Park Zoo, and by the end of 1936 both Staten Island and Brooklyn had new zoos as well. A third figure, former New York Governor Al Smith, who was given a symbolic position at each new zoo, played the part of cheerleader, entertaining the press with his frequent and amusing stories about the animals he loved.

The construction of all three zoos provided New York's political and civic leaders with compelling evidence of how New Deal policy could benefit average citizens. Mayor LaGuardia recognized zoos' enormous popularity and welcomed them as free, wholesome entertainment for a public he sought to please. The advocacy of politically well-connected civic leaders, such as Iphigene Sulzberger, the wife of the *New York Times* publisher, gave LaGuardia and Park Commissioner Moses additional reasons to embrace zoo construction. Perceiving the utility of New Deal funds for his broader program of reshaping the city, Moses used the Central Park Zoo project to demonstrate his efficient and visionary planning talents.[54]

The Central Park Zoo was among Robert Moses's first grand successes in transforming New York's public landscape. It took on special significance as a symbol of the efficiency that he brought to the task of building public works. In a broader sense, the zoo's rapid construction and pleasing design also illustrated that relief laborers were not without skills. And, its immense popularity immunized the project from any charges that the New Deal was not serving the needs of the public.

In the decade prior to Moses's arrival, the Parks Department had made some efforts to improve the zoo, but they had been sporadic and uncoordinated, the onset of the Great Depression curtailed significant park work, and political support for the zoo was lukewarm at best. During the 1920s, the *New*

York Times editorialized several times that the dilapidated menagerie blighted Central Park and should be removed altogether. In response to such threats, the Department of Parks *Annual Report* for 1929 reported on several projects: an enlargement of the seal tank, the construction of a "winter stone hut," minor repairs and painting of building interiors and exteriors, and the installation of electric blowers "for the purpose of disinfecting and ventilating the Lion House." But by 1930, the zoo's annual budget was only $20,280, with no funds spent on physical improvements. A 1931 Regional Plan of New York and Its Environs recommended removing the zoo to Wards Island and demolishing the adjacent Arsenal building, in which the Parks Department was headquartered.[55] Utilizing New Deal resources, Moses swept aside such suggestions, announcing in early 1934 that he would expend $411,000 to construct an entirely new zoo and refurbish the Arsenal building.

Moses selected Aymar Embury II as the consulting architect of the Park Department, and he chose Embury to personally design the new Central Park Zoo, which he considered to be "his best work." Embury was among the highly skilled and well-compensated (he was paid $50 a day for his Park work) employees that enabled Moses to manage and complete so many projects at once. Embury's architectural style, which Moses later described as "invulnerable to changing moods and durable physically and aesthetically," proved well suited to the needs of an urban zoo. According to Moses, the new zoo buildings were "of no particular architectural style, but are perhaps a modernized version of Victorian architecture characteristic of Central Park, simplified to the last degree." Perhaps as impressive to New Yorkers, however, was the fact that it took only 16 days to design the zoo and eight months to build it.[56]

Moses could build the zoo so quickly in no small part because the New Deal relief programs emphasized mass employment, meaning that many hands, however unskilled, were available for labor. Using Civil Works Administration funds, Moses assembled a large team of engineers and draughtsmen; in addition to Embury, 15 men worked on the zoo's design plans, and Moses inspired or bullied many to work beyond their paid hours. The actual construction utilized a veritable army of laborers working literally around the clock. Nearly 700 men worked a day shift, and then another 400 took over for a night shift made possible by high-powered carbide lights.[57]

The zoo's opening ceremonies, only eight months after construction had begun, were filled with pageantry and praise. Former Governor Alfred Smith, escorted to the stage by 300 children, reminded the crowd of 15,000 that Moses had completed the zoo in record time, and he praised it as "one of the finest of its kind in the world." The zoo's immense popularity was clearly enhanced by Moses's appointment of Smith as the "honorary night superintendent." With a key to the cages, Smith was charged with keeping an eye on the zoo and locating animals to fill the new cages. Smith, who had lost the Democratic presidential nomination in 1932, was a brilliant choice for this role. First, he genuinely loved animals, having had a small private zoo on the grounds of the Governor's mansion and keeping several pets in his Fifth Avenue home. Second, the press loved Smith, who was quick with a joke. And finally, Smith's animosity towards FDR and his public criticisms of the New Deal and its "Boon-Doggle" projects (he kept a scrapbook full of clippings about these) meant that his positive affiliation with Moses's park projects would ensure that they were seen as legitimate. Smith took to this new role with glee, visiting the zoo, reporters in tow, to feed and show off his favorite animals. Ever a loyal Democrat, he cracked jokes at the elephants' expense and trained a tiger (the Tammany Hall symbol) to roar whenever he yelled "La Guardia." The press reported fairly regularly on Smith's zoo antics, generating positive publicity. A year after the zoo opened, Moses wrote Smith to say that "the Zoo business is getting along very well, thanks to your new animals."[58]

The rapid construction of an architectural gem cemented Moses' reputation as an adroit bureaucrat able to coordinate support from numerous alphabet agencies to implement a grand vision of construction for the city. When he compiled a six-year report for Mayor LaGuardia, Moses cited the Central Park Zoo project as "the first indication of the efficiency of the new staff of professional designers." Even someone as close to the process as Aymar Embury expressed private admiration for Moses' "extraordinary imagination" and "enormous grasp of detail" when he realized just how extensive his boss's projects were. Moses modestly denied that there was any "trick formula" to his success; rather, "it comes down to having a lively interest in a program, some knowledge of the working of government machinery, and a sufficient determination to see the program through."[59] Next up for Moses was the Prospect Park Zoo project.

For over 30 years Brooklyn's civic leaders had been trying to modernize their humble menagerie, but it took New Deal funds to realize their ambitions. The Central Park Zoo had been built quickly to demonstrate what Moses could achieve if given free reign. After only two months in his job, Moses announced his intention to give Prospect Park a real zoo, its shape to be guided by the Brooklyn Citizens' Committee for the Prospect Park Zoo.[60] This committee was headed by Brooklyn Borough President Raymond Ingersoll and by Moses. It coordinated with the Brooklyn Chamber of Commerce to generate public support for the zoo and raise money to buy new animals.[61] Alfred Smith's role as zoo cheerleader took on greater significance in this project, and the completed zoo further demonstrated the benefits of New Deal public works.

Brooklyn's Prospect Park, like many city parks across the nation, included some land set aside for a "menagerie" that had taken shape in the late 19th century and slowly evolved into something resembling a zoo, with a few permanent buildings and many make-shift enclosures and cages. The menagerie collection in 1894 ranged from "one sacred cow" to six bears and 16 rabbits. By 1915, civic-minded citizens who were embarrassed by a zoo that "had no Lions or other representatives of the Cat family" raised $3,000 to buy some 90 new animals. Joining together as the Brooklyn Zoological Society, these same citizens raised another $16,000 to build an "attractive two-story brick structure" so that "the public will then have its first good opportunity to see our collection" properly displayed. Under this system of improvement, private citizens paid for the physical improvements, thereby "costing the City no more for maintenance than was formerly the case. This claim proved unfounded, and by 1923 Park Commissioner John Harmon called the zoo a "disgrace" filled with "old and decrepit" animals living in "badly planned" or "wholly inadequate" structures. No public money, however, was available to alleviate these conditions, so Harmon appealed to Brooklyn's residents to send donations to the Zoological Society so that it could buy better animals and construct new enclosures. Over the next few years, the zoo gained a bear cage, a new elephant house, and several paddocks for various animals. When Moses turned his attention to Prospect Park, he found its zoo buildings "in little better condition than those in Central Park," and he had most torn down to make way for a grand New Deal zoo designed by Embury.[62]

When the zoo opened on July 3, 1935, LaGuardia and Moses were absent because of a minor political feud, so Smith was the de-facto master of ceremonies. After his speech, Smith enjoyed a beer at the zoo's restaurant, and then led reporters on a tour of the new zoo, which was described by the Parks Department as "one of the most modern and beautiful menageries in the country." Comprised of six buildings arranged in a semicircle around a seal pool and landscaped grounds, the zoo combined indoor and outdoor exhibits. Some of the outdoor exhibits were "modern" in the sense that they separated the public from the animals with a deep moat and low wall, offering a bar-free view of elephants and bears. As in the Central Park Zoo, artists from the WPA sculpture and painting divisions made the Prospect Park Zoo a showcase for architectural adornment. Bas reliefs, sculptural elements, and murals were integrated throughout the buildings, gracing interior and exterior walls alike. Al Smith found an opportunity for political commentary in the art, observing that F.G.R. Roth's elephant heads were too partisan and should be balanced by "some donkeys." All in all, the architecture and landscaping were quite pleasing, and Moses described the Prospect Park Zoo as among "the most important buildings" Embury designed. And, like the Central Park Zoo, it created political capital for him, eliciting "much outspoken praise" from Brooklyn civic associations, which would be important to him as he advanced his ambitious road, bridge, and park construction plans over the next few years. Despite Moses' apparent aversion to dedicatory plaques, the zoo would recognize the private citizens who had purchased the new animals with a bronze marker in front of each exhibit.[63]

Although conceived of and promoted by a maverick park commissioner, the Central and Prospect Park zoos were very similar to many other city zoos of the era in the sense that they were managed not by a private zoological society, but by a Parks Department with limited expertise and understanding of wild animals. The zoos shared a director, Ronald Cheyene Stout, who perceived zoos' central public purpose to be recreational entertainment and took every opportunity to play the clown. Stout enjoyed entering the cages (sometimes at night) while occupied to demonstrate his fearlessness — he claimed that cast-off circus lions were "sissies" — and the papers reported gleefully on the lost fingers and assorted maulings that occasionally resulted.[64] Alfred Smith attracted similar attention, gathering photographs and headlines when he would invite reporters along during his (supposed) nightly tours of the zoo in his official capacity as night superintendent.[65] Smith's visits were full of jokes and always demonstrated that he knew each animal by name. The animal collections themselves emphasized the display of quantity and familiarity over any scientific purpose. Lions, tigers, bears, chimpanzees, elk, and elephants populated the cages, many donated by circuses, pet owners, or governors. Both zoos attracted huge crowds and clearly satisfied a public desire to be entertained by animals. Those citizens who sought knowledge from their zoo experience had to visit the Bronx Zoo or wait until the Staten Island project was completed.

The Staten Island Zoo was the third of Moses's zoo projects, announced along with the others in February of 1934.[66] Citizens expressed strong support for the proposed zoo's educational mission and for the use of relief funds to build the zoo. Soon after Moses formally announced in March of 1934 that the project would move forward, one letter to the editor from a nature enthusiast stressed that the zoo would be "a valuable asset to the educational facilities of Staten Island" and welcomed the unemployment relief to be provided by its construction. Another letter writer praised Moses and argued that "no CWA funds could be better expended than for this building which will provide just those recreational facilities — for adults as well as for boys and girls — which President Roosevelt and other leaders of today have strongly indicated the future will require."[67]

Despite the Staten Island Zoo Society's (SIZS) success in generating public support for a new zoo, the Parks Department did not complete the project with the speed promised. Staten Island residents attributed construction delays to Parks Department foot-dragging, but it seems that the inefficiencies associated with building a zoo under an evolving set of relief programs also contributed to the delays. Work on the zoo began in 1933 as part of the city's work relief program, but was then transferred to the Civil Works Administration in late 1933, with only a portion of the foundation excavations completed. Moreover, Park Commissioner John J. O'Rourke opposed the zoo project altogether, arguing that "the people have been trying to get rid of [the Central Park Zoo] for 20 years." O'Rourke used bureaucratic maneuvering to stall the project, arguing that the Art Commission and the Park Board's landscape architect had not yet approved the building plans. He explained that he had allowed CWA workers to do a little foundation work just to keep them busy while trying to assign them to other jobs, and he indicated that he had discouraged the CWA administrators from supporting the zoo project. In response, SIZS member Harold J. O'Connell remarked, "there is no doubt that Commissioner O'Rourke has been giving us a run-around."[68]

When Robert Moses took over as Park Commissioner, it appeared that the necessary political support for the zoo was finally in place. An official Parks Department press release in March 1934 promised that CWA workers would break ground "immediately" and that "the Zoo will be finished and ready for occupancy by early summer."[69] By June Moses had pushed the completion date back to early November. On June 4, 1935, Moses promised the Staten Island Chamber of Commerce that although "a reduction of work relief funds" had stalled projects in Richmond Borough, they would begin again after July 1. Two weeks later, the $300,000 work-in-progress was inspected by a large group of local and state officials on a tour of Staten Island's park projects. Unsatisfied by these promises and inspections, the *Staten Island Advance* warned that unless construction was completed by September, the zoo's growing collection of animals would continue to be housed in an unsafe two-story barn, where visitors could not view them. Noting that the Central and Prospect Park zoos had already been finished, the paper editorialized that "this looks like out-and-out discrimination" by Moses against Staten Island and the Richmond Borough. Construction slowed again in the fall when 300 WPA workers quit in protest over delayed paychecks.[70]

With the zoo still unfinished in March 1936, the *Staten Island Advance* editorialized that the project was "No Advertisement for Mr. Moses." The paper continued to attribute the delays to the low priority Moses assigned the zoo and noted that a private contractor could have completed the zoo "in a month." Moses replied not to the paper but to Harold O'Connell, the Staten Island Zoological Society vice president, chastising him for not contradicting the editorial's "grotesque statements" and pointing out that he had long argued for contract relief precisely to avoid the labor inefficiencies that had slowed the Staten Island Zoo project. The slow pace of progress on the zoo was due, Moses claimed, to the various deficiencies of relief work. By late spring of 1936, Moses increased the zoo's WPA force from 300 to 1,000 men, working two shifts (from 8 am to midnight), seven days a week.[71]

In fact, the zoo would not open until June 1936, but the finished project perfectly served the SIZS's goals. Located on an eight-acre tract, the modest T-shaped structure of red brick and limestone featured "glass shingles" to allow natural light through the roofs into the exhibits, a modern reptile wing with an alligator pit, as well as cages decorated with WPA-painted murals that replicated the animals' native habitats. Press reports emphasized the zoo's accessibility to New York, being only 15 minutes by bus or car from the Staten Island ferry terminal. Advertised as the "most modern zoo in the country" (a claim seemingly made for many WPA-built zoos), the zoo's opening ceremonies were held on June 10, 1936, and attended by 5,000 people, including Robert Moses, Alfred E. Smith, Staten Island Borough President Joseph Palma, and SIZS member Harold J. O'Connell.[72] The speeches given during the opening ceremony offered the usual praise for zoos in general and Moses in particular.

In less than three years then, New York City added three entirely new public zoos, thanks to the largess of the New Deal. Moses had worked quickly, fearful that federal funds might disappear at any moment, but other cities proceeded at a more leisurely pace. In New Orleans, local park officials were about a year behind Moses, but FDR was re-elected in 1936 and the New Deal continued to build zoos.

The Audubon Park Zoo

New Orleans was one of the most economically significant cities in the nation, a shipping and financial center that was New York City's Southern rival. Lacking harsh winters and enjoying easy access to traders from around the globe, it seemed like a perfect place for a zoo, one as grand and flamboyant as the city's history and culture. Indeed, Audubon Park, located on the eastern curve of the Mississippi River just northwest of the downtown, was home to the Audubon Zoo, over which the New Orleans Zoological Society and the Audubon Park Commission doted hopefully. But this zoo did not live up to the city's self image. Instead it occupied the vast third tier of American zoos — institutions undistinguished in collections, design, quality, or leadership. As the roaring twenties wound down, New Orleans' leaders faced the future fairly confident of the zoo's survival, but with no sure plan about how to elevate its quality. A $50,000 bequeathal in 1929 promised to see the zoo through any rough times. Few at the time could have imagined that a decade later that gift would have been leveraged into zoo improvements worth nearly ten times as much.

The Audubon Zoo had a typical origins story. It was the gift of wealthy patrons who controlled the John Charles Olmstead-designed Audubon Park for public benefit. The park hosted a small, informal menagerie until 1913, when the Commission dispensed with it on the grounds that a zoo was incompatible with the park's Olmstead aesthetic. Just three years later, however, a flight cage designed by William Hornaday was erected. Daniel Moore, the manager-editor of the *Times-Picayune* and president of the Audubon Park Commission, gave $500 in seed money to start the zoo in April 1919. A New Orleans Zoological Society was created to support the effort. Over the next decade the zoo added animals and cages, growing into a popular local destination. Then, in 1929, just three months before the stock market crashed, Valentine Merz bequested nearly $50,000 to use for the zoo's development and maintenance.[73]

The Audubon Zoo was a case in which a private group of elite citizens controlled the management of Audubon Park and its zoo for public use, but lacked the financial resources to develop the park in the ways they wished. The city provided a small annual appropriation of around $10,000 to support the zoo, but this sum was inadequate for the Commission's purposes. A proposed property tax to benefit the zoo had been defeated by voters in 1928; only property owners could legally vote on such a referendum. Aside from the Merz legacy and the occasional special purpose donation, the Commission did not have a substantial amount of private capital at its disposal. As the Depression began to cast its shadow, it looked as though the zoo would never be able to grow beyond its present size, and perhaps it might not survive at all. But, when relief for the unemployed reached Louisiana in the form of FDR's New Deal, the zoo's fortune changed.

The Audubon Park Commission did not make immediate use of relief labor programs to benefit the zoo, but when it did begin to do so, it moved quickly and aggressively. Between May 1935 and October 1935, the commission contributed $2,485.39 to enable FERA laborers to repair and strengthen the existing zoo's cages, putting them in "good, safe condition."[74] Even the lion house, which had been built in 1927 and was advertised as "the latest word in design, construction, and adaptability for animal cages," needed work.[75] These repairs were clearly valuable, but they did not fundamentally change the stature of the zoo. Moreover, the commission could not be sure how long the "free labor" would be available. Recognizing the necessity of moving quickly, the commission simultaneously used $1,000 from the Merz Fund to have James Dawson, a landscape architect with the Olmstead Brothers firm, visit New Orleans "without delay" and prepare a preliminary plan for having the zoo moved to the southwest end of the park.[76] Frederick Law Olmstead had designed the original Audubon Park, and the commission sought a redevelopment plan that would be consistent with his vision. The commission was being both conservative and aggressive in its actions. Committing some money to the FERA improvement of the existing zoo was a good hedge against the $1,000 bet that with a set of new zoo plans in hand, the commission would be able to get authorization for a much more ambitious zoo project. Discussions about such an expansion had begun in 1933, apparently fueled by the sudden availability of federal funds for public works projects. Now, just two years later, the commission was poised to act.

The Olmstead report outlined a vision for a new zoo that would improve both the zoo and the park. Dawson's plan called for moving and rearranging the existing zoo buildings to give the animals more space and the zoo itself a unified and attractive design. Reflecting his firm's fundamental bias against zoos and distinct disapproval of the old site, Dawson argued that the old zoo space would be "more valuable" for other purposes anyway. Nonetheless, Dawson expressed some familiarity with innovations in exhibit design, and he recommended the construction of a few "barless cages" where visitors were separated from the animals by a moat rather than by bars. New Deal projects in St. Louis, Detroit, and San Diego had already constructed such exhibits, which were expensive but had been made affordable by "welfare labor." Interestingly, Dawson preferred barless cages not because they improved the animals' welfare, but because they eliminated the "objectionable features of looking through bars and wire fences." Aesthetics guided Dawson's plan, and he spent most of his report detailing the landscaping features he imagined for the new zoo.[77]

However aesthetically highbrow, the planned zoo move put the commission in conflict with the Park's well-to-do neighbors. The Garden District around the park was distinguished by grand homes and residents who sought to protect their property values. A park along the lines of the original Olmstead design, which favored natural scenery and quiet relaxation over recreational facilities was fine with these residents. The park's Bridle Club represented the kind of recreation that these better citizens wished to see in the park. When news of early (1933) discussions to move and expand the zoo reached these citizens, they responded with disapproval. Communicating through his attorney, one neighbor warned that he would be "forced to protect his rights should the occasion arise," and others wrote the park superintendent to "voice their vigorous, emphatic and unanimous protest against this purported move." They feared that "incessant noises" and "noxious odors" would lessen the value of their homes.[78] After the commission had already applied for and received WPA funds to move the zoo, a Bridle Club representative approached the commission to plead against any encroachment on its area.[79] But the spirit of the New Deal was inexorable, the temptation of subsidized expansion irresistible, and the motivation of civic pride too powerful. The Board of Commissioners, debating genteelly in the Hibernia Bank boardroom, sided with the elephants, monkeys, and lions. New Orleans would have a proper "Zoological Garden," one whose quality would properly reflect the city's status. Civic pride was at stake.

Moise Goldstein, an architect and chairman of the commission's Ground Committee, shepherded the zoo project to its completion. Working loosely from the Olmstead plans, he drafted detailed architectural blueprints and worked closely with the park engineer to supervise the use of the relief labor. Goldstein, like Moses in New York, fully appreciated the necessity of working quickly to use New Deal funds before they evaporated. Even before he had prepared detailed construction blueprints, he had survey crews staking out the main axes and paths at the new site. Two months into the project, the WPA work crews were working two separate shifts, slowed only by a shortage of skilled labor. Goldstein happily reported on the "well balanced plan," which would transform the old zoo into a "place of natural beauty."[80]

Plans and costs for the new zoo soon expanded. The original plan submitted for project approval called for wood frame structures of the same style found in the existing zoo. Soon after the project received approval, however, Goldstein recommended that the new zoo be of more permanent brick construction, and he drew the blueprints accordingly. The new zoo grew in size too as the commission applied for WPA "supplements" as additional funds became available. Thus, the original project, which began on October 23, 1935 (just after FERA repair work on the old zoo was completed), cost $253,000; two supplements, received in April and July of 1936, brought the total cost to $336,000, of which approximately $42,000 had come from the commission, drawn mainly from the Merz Fund. This first phase of construction concluded in February 1937, whereupon the commission sought a second major project worth $117,000 to build a new elephant house, a new lion house, and a restaurant, and to landscape the new zoo. The total cost of the new zoo when it opened in 1938 was close to a half million dollars. Originally, the commission had estimated that by committing some of the Merz Fund to relief labor projects, the park "could accomplish about three times as much work."[81] In fact, thanks to the New Deal, the Merz Fund had been leveraged into nearly ten times its original value.

The new Audubon Zoo (or Merz Memorial Zoo, as it was officially named) opened to great fanfare and much praise, but relief laborers remained in the zoo through 1941. Thanks to the WPA's arts division, John McCrady and other artists carved animal reliefs into the brick faces on several bird and animal houses.[82] These artistic adornments, which cost the commission nothing since the "materials" were already in place, had been suggested by Goldstein, and they further enhanced the new zoo's physical attractiveness. Other WPA projects allowed for the "reshaping" of the brand-new Monkey Island, which was already eroding around the edges, some alterations to the bird house, and an enlargement of the popular Snake House. Some repairs were necessary for safety reasons. Two Kodiak bear cubs given by Governor Leche had grown so quickly that their cage needed to be strengthened to hold them. Other projects addressed more mundane but vital needs: in 1941 WPA crews returned to repair the zoo's inadequate sewage system to avoid any "unpleasant criticism" that might have resulted from its failure.[83]

State and local political support for New Deal public works projects had been another important factor in the Commission's ability to secure WPA projects. Although Governor Huey Long's relationship with FDR turned adversarial after 1932, Long was a strong proponent of public works projects. New Orleans mayor Robert Maestri shared Long's political philosophy. In case anyone had any doubts of this, he proclaimed his undying support for Long in an ad he placed in *The American* (one of New Orleans' newspapers) that said in large, bold letters: "Huey — I have tried my best to follow in your footsteps — you are constantly in my mind and every move is directed along suggestions you made to me in days gone by." Most important for Long, and therefore Maestri, was the basic infrastructure development that the Louisiana elite had so long denied their state. New Orleans was severely lacking in paved roads, for example. James H. Crutcher, the Works Progress administrator for Louisiana, praised Maestri for "the many accomplishments which have resulted from your complete co-ordination with the WPA." Foremost among those were the "street paving program, which, with the assistance of the WPA has brought about the improvement of more than twenty miles of thoroughfare." However, he also praised "the remodeling of Audubon park zoo, and scores of lesser projects which have provided work for thousands of the city's unemployed." He praised Maestri's similar "theories of caring for the indigent [that] have coincided completely with those of the national administration." He reminded the public that Maestri believed that "men and women who, through no fault of their own, have temporarily lost their place in private industry ... must keep their self-respect," which could only be maintained through" worthwhile occupation." Citizens in New Orleans thought of themselves as "not on relief, but engaged in a worthy task for the general welfare of their city." Mayor Maestri received credit in the local newspapers for his "skillful ministration" in securing WPA funding for the zoo.[84]

The New York and New Orleans zoos were significant, architect-designed institutions worthy of national attention. They were not the largest or the most beautiful, but they represented what the New Deal made possible. Scores of zoos across the United States received new buildings and exhibits thanks to federal relief agencies. Not every zoo benefited on an equal scale, of course. The Central Park Zoo was known as a "postage stamp zoo" because of its tightly compressed space, but in comparison to some of the modest zoos built elsewhere by relief labor, it was a like a gaudy, oversized, airmail stamp. The Washington Park Zoo in northern Indiana, for example, nestled into a few acres of a park on Lake Michigan, represented a more modest version of what a New Deal zoo could be. Here was a humble, charming structure that delighted local residents just as much as the big city zoos delighted theirs. The majority of New Deal zoo projects were in small communities like this and were decidedly local in their scope and ambition. But, like other New Deal projects, zoo improvements were spread more or less evenly across the American landscape.

West, east, north, and south, America's cities either had zoos or desired them, and local zoo directors, politicians, and citizens greeted the New Deal zoos with civic pride. After the Audubon Zoo's animals were moved to their new WPA-built facility, the park superintendent enthused that "the development of the new Zoo has brought to us the finest zoo in the country. Our buildings have been admired and favorably commented on by all who have visited our zoo."[85] Some version of this comment was expressed in every zoo touched by the New Deal.

The zoo landscape was transformed by the New Deal, but why? The public value of other public works projects — dams, roads, schools, post offices — was fairly self-evident. But how did zoos serve the public interest? What business did the state have in providing Americans free access to captive animals? Were there not other more pressing needs to be satisfied first?

TWO

Who Ran the Zoo?

"Dear Mrs. Mann,

I just learned of the passing of your distinguished husband. As one who knew him for many years and admired him for a much longer period, I cannot refrain from tendering sympathies and best wishes. For 25 years I served in the U.S. Fish and Wildlife Service. On a number of occasions I was in your home and I saw Dr. Mann on innumerable occasions out at the zoo. Many times I took ants out for him to identify and to give me information concerning them. I am quite sure that his help to me was about as it was to a great many other government employees and men interested in science in all parts of the country and in foreign lands ... I am sure you and he had long forgotten it but when I came to Washington as a Junior Biologist with a young wife and three youngsters I received a call from him one day to attend a circus. He assured me that the tickets were already prepared as he knew I was a Junior Biologist and, therefore, he had paid for our tickets or had obtained passes for us. I know others whom he treated similarly. He was the greatest zoo leader that America has had, in my opinion. Consequently, the entire nation is indebted to him."

Clarence Cottam, Director, Welder Wildlife Foundation, Sinton[1]

"This is a loss which will be felt personally and professionally by literally thousands and thousands of people, which is not very much consolation to you, who must feel it acutely."

C. W. Coats, Director of the New York Aquarium[2]

> "Dear Sir—
>
> Mr. John McLaren, the well-known Park Superintendent of San Francisco, has often suggested the desirability of San Francisco's having a zoological garden—something it has never had, although there are some fine specimens of ruminants in Golden Gate Park, and we have a considerable aviary there. The mild and equable climate here should be ideal for the purpose.
> In order to go further into the matter I should like to get as full information as possible about how zoological parks are handled elsewhere. I should be greatly obliged for any information on such points....."
>
> *W. M. Storther, Secretary of the Board of Trustees de Young Museum*[3]

In their letters to Lucile Mann, C. W. Coats and Clarence Cottam captured the great affection and respect that many Americans felt for Dr. William M. Mann, the long-time director of Washington, D.C.'s National Zoo who died in 1960 after a long, painful struggle with arthritis and a weakened heart brought on by malaria.[4] Confinement to a wheelchair during his final years must have been particularly hard on a man who had traveled so extensively throughout the world and changed the face of American zoos after becoming the National Zoo's director in 1925. Mann had entered the zoo profession after already establishing an international reputation as an entomologist, and as W. M. Storther's request suggests, his academic credentials and his position at the National Zoo made him one of the nation's central authorities on zoos. The circle of young naturalists who visited and consulted with him regularly included Marlin Perkins, later the director of the Buffalo and Lincoln Park zoos, and Roger Conant, the director of the Philadelphia zoo, among others.

William (or Bill, as he was known to friends) Mann was part of a vanguard of young men, many with an academic interest in herpetology, who began their zoo careers during the 1920s and 1930s and played leading roles in sustaining and transforming this public institution. Like Mann, these men had at least some formal university training, and through their zoo positions they conducted research and participated in a range of scientific and conservationist activities. Together, they influenced the development of American zoos, both by working through a new (1924) professional organization—the American Association of Zoological Parks and Aquariums (AAZPA)—to promote animal exhibition and care standards, and by choosing and mentoring staff. Individually, and sometimes collaboratively, they advocated a science education mission for zoos, supported wildlife conservation efforts, and defended the theory of natural selection. Under their leadership, zoos grew in size and scope, even during the depths of the Depression.

Although Mann and his peers spoke a language of science understood by few, and they sometimes enjoyed social ties with the power elite of their communities, there was little snobbery about them. On the contrary, most felt a great affinity with and affection for other lovers of animals. Thus, they welcomed the opportunity to popularize their knowledge, and they socialized easily with circus performers and others in the exotic animal business. Their passions were the study, collection, and exhibition of exotic animals, and those passions enabled them to find common ground with Americans of all ages and from all walks of life. With the help of New Deal funding, they made American zoos both more scientific and more entertaining.

William (Bill) Mann, Zoo Man for an Era

Like many other zoo workers in the 1930s, Mann grew up in a middle class home and was an amateur naturalist from his earliest days. He was born in Helena, Montana, in 1886 to parents who had settled there after arriving from Missouri by covered wagon. His mother, Anna, was an immigrant from Wales who came to the United States as a young child. A devout Baptist, she read the Bible frequently in both English and Welsh. His father, William M. Mann, a harness maker and amateur inventor and taxidermist, died in 1893, leaving Anna to fend for her only son during a national depression. Luckily, the family owned two houses, so Anna sold one, rented the other, and moved into smaller quarters. Rent money and occasional checks from a Welsh aunt in Dodgeville, Wisconsin, sustained the two through difficult times. Willie (as his mother called him) also had a half-sister named Lily and a half-brother, Charlie, who lived in Chicago.[5]

Willie was an adventurous, impulsive child, who dreamed of joining the circus but actually ran away to a ranch and stayed there for six months. Willie's adventure began when Mr. Beinhorn, his seventh grade teacher, quit to pursue his lifelong dream of cattle farming. To Mann's great delight, Beinhorn offered him the job of cow herder, which effectively meant that Mann could pursue his growing interest in the natural world while the cows grazed. Mann lived and worked on the farm during the summer of his seventh grade year. One day, however, Mann found a skunk that distracted him so much that he failed to bring in the cows on time for milking. Mr. Beinhorn beat him for his transgression. Humiliated and angry, Mann collected his small number of possessions and left the Beinhorn farm. He followed the road to a mining camp at Marysville where he explained that he was new to the area and had wandered too far and gotten lost. He repeated this lie each day to families that generously took him in as he fled further from the farm. Clearly, Mann was not heading home.

Eventually Mann found a rancher named Ed Chaple with a small herd of cattle in Warm Spring Creek who was willing to take him on as a hand. Mann explained that he needed a job because he had been treated meanly at a circus and had run away. He claimed that his job at the circus was training horses. He even adopted the name Toby Tyler, the protagonist from *Toby Tyler, or Ten Weeks with the Circus*. Chaple later revealed that Mann's ineptitude with horses made him doubt that part of the story, but the rest he accepted. Mann settled in on Chaple's farm and remained there without informing either his mother or Beinhorn of his whereabouts. Initially neither party had any concern because each imagined him at the other residence. Anna thought that he was still working at Beinhorn's, and Beinhorn believed that he had gone home to his family. But when Mann failed to appear for the fall school term Anna got worried and took out ads in several Montana papers, including the Butte *Weekly Miner* where Chaple read her missing son notice. Chaple confronted Mann, asking him, "Say, young man, ain't your name Willie Mann?" Soon Willie was back with his mother. In his autobiography Mann claims that it never occurred to him that she would be worried about him, suggesting a range of interesting questions about their relationship, but it seems that his desire for adventure overwhelmed his need to be physically close to her.[6]

Ironically, Mann's high school education took place at boarding schools some distance from his mother. For the eighth grade, he attended the Lyon's boarding school in Spokane, Washington. Here Mann was exposed to a range of religions and interesting characters who broadened his horizons. Each Sunday, for example, Mr. Lyon took the students to a different church. Students attended theater, heard talks by visiting missionaries, and met a range of teachers at the school who stayed for a while and shared their experiences. One teacher was a preacher who had spent time in Brazil and owned a pet monkey and parrot. Mann enjoyed the school and the outdoor opportunities it afforded, writing his mother to describe ten-mile hikes and interesting specimens that he had collected. The Lyon's school ended at the eighth grade, however, so Mann had to leave again. At Chaple's suggestion, he went to a military school in Staunton, Virginia, where he wore a uniform, rose at 5 A.M. to study, and found time to appreciate the animal life in the local park.[7] For reasons that are unclear, Mann left this school after a year to live with his half sister Lily and half brother Charlie in Chicago. There he attended the Lewis Technical Institute only to fail at Mathematics, German, and Latin. Because he had done so poorly, the family sent him back to the military academy, where he completed his high school education.

Despite a somewhat checkered education, Mann's interest in natural science helped him gain admission to college. By the time he graduated from high school, Mann had independently learned a great deal about entomology, in part by becoming a studious amateur collector. In the absence of standardized admissions tests, early twentieth-century professors actively recruited students gifted in their particular expertise. Mann was one of those students. He was part of a very small world of amateur animal collectors that included professors and university presidents. Mann's acquaintance with the professorial world began when he would lack identifying information about the insects. To learn their correct names he corresponded with college professors, sometimes offering them the insects for their collections. For example, before attending college he corresponded with a Professor Wickham at the University of Iowa and sent him specimens of striped weevil.[8] Thus, once past secondary school, Mann's ticket into higher education was his collecting zeal and detailed knowledge of insects, both of which were quite useful to scientists researching how to manage agricultural pests.

For his undergraduate education, Mann chose to attend the State College of Washington because he had read an article about insects by one of its faculty, Professor Rennie Wilbur Doan. Mann brought his collection of insects to Professor Doan only to find that he had left for Stanford. So, Mann left his collection with an aspiring entomologist, a lab assistant named Rueben Trumbull, who gave them to the entomology professor, A. L. Melander. When Mann went to register for both Zoology One and Entomology One his first semester, he learned that the first course was the prerequisite for the second. He tried pleading with Melander to admit him without the prerequisite, but was treated as a typical student and reminded about the rules of the university. However, when Melander learned that the recently-donated insect collection was Mann's, he waived the prerequisite.[9] The rules were easily bent for students who could help professors with their research, and Mann pursued entomology with a passion. In addition to his coursework, he corresponded with William Morton Wheeler, the Harvard University authority on ants and one of his heroes. Mann dreamed of someday studying under Wheeler.

College professors like Melander and Wheeler were taxonomists for an entomological world that also included government scientists at the Bureau of Entomology in the Department of Agriculture. Wheeler and his fellow entomologists catalogued each newly discovered species with exacting detail, producing drawings and descriptions that would assure identification by other scientists.[10] Such identifications were useful not just to amateur collectors, but also to those battling economically destructive ants and other insects. Melander and Mann, for example, helped apple growers in Washington who were struggling against the coddling moth, by redesigning an insecticide sprayer to treat the top of the apples where the moths lay their eggs. As a byproduct of their field research, Mann identified additional new insect species.[11]

The shortage of graduate students meant greater research opportunities for undergraduates. In addition to assisting Melander, Mann worked at the Puget Sound Marine Biological Laboratory and collected research specimens for the college. He also met Stanford ichthyologist E. C. Starks, who persuaded Mann to transfer to Stanford with the promise of a research assistantship under the head of the zoology department, Vernon Kellogg, formerly the state entomologist of Kansas. Thus, in his senior year Mann was drawing a 25-dollars-a-month salary, enough to hire his own research assistant, Lee Dice, who later became a professor at the University of Michigan. At Stanford, Mann discovered another exciting opportunity: an expedition to Brazil directed by John C. Branner, professor of geology. Branner had already staffed the expedition with leading faculty from several universities, including Starks. Mann pleaded with Starks to be added as an entomologist, and Branner agreed, on the condition that he raise his own travel funds. Mann wrote to private collectors and natural history museums, promising to collect on their behalf, and he soon had enough money.[12]

Mann's 1911 expedition to Brazil helped secure his career in science. Mann recognized its significance even before departing, writing to his mother ("the best mother a boy ever had") to thank her for sending some money. He claimed that landing a position with this expedition was "the really great success of my life so far," since he would be "with the greatest men in their lines in the world." Their importance was acknowledged by Theodore Roosevelt, who met with the expedition's members before they left. In another letter written while traveling on the Amazon River, Mann gushed with enthusiasm about this "delightful trip" and the amazing variety of wildlife that he was seeing. More newsworthy was an offer from the National Museum of Brazil to head up its entomology division, a temptation topped only by Professor Wheeler's invitation to become his graduate assistant at Harvard. Wheeler promised a $10 per week salary with no duties other than studying, as well as a free room in his own home. Mann would return to America with a world-class collection of Brazilian insects, and he vowed to his mother that he would "make something of myself yet for your sake."[13] He indicated his maturing status by signing his letters "Will" instead of "Willie."

In the fall of 1911, Mann settled in at Harvard, quickly getting to know the faculty and throwing himself into the "work I have so long wanted to do." By publishing a paper on the ant collection from the Brazil trip, Mann knew that he would become known as "something besides a collector." Still, he did not want to depend on his scientific reputation alone, and he vowed to "dress decently" because his "entire future depends on what the people here think of me." Indeed, his Harvard professors thought quite well of him, becoming his close friends, enabling him to travel the world on collecting expeditions, and granting him a biology doctorate degree in 1915.[14] Once a fully degreed member of an elite scientific community, Mann became known as "Bill," and his independence grew. After graduation, he went on a two-year collecting trip around the world; on this expedition he contracted the malaria that plagued him for the rest of his life, eventually leaving him in a wheelchair with severe spinal arthritis and severely weakened heart muscles.

Upon returning to America, Mann looked for a new position. On his way to a job collecting hummingbird moths in the Bahamas, Mann stopped in Washington, D.C., to visit the National Zoo and friends in the Department of Agriculture's entomology section. While there, Dr. Marlatt, the chief of the department's Federal Quarantine Board, offered him a job working for the government as an ant specialist and collector in the tropics. Mann accepted, becoming an associate curator of insects. In a typical year he would spend six months in what the Bureau referred to as the tropics and six months at home analyzing his collection. Mann traveled to Europe as well, visiting Spain a few times, for example, because the Bureau worried about the importation of fruit flies that were plaguing the Valencian orange growers. The job paid $1,800 per year, and Mann liked his work.[15]

Mann's collecting tastes began to broaden, and his career took a twist. In addition to collecting insects, he started to pick up rare animals that he knew would be of interest to colleagues in other disciplines or to his contacts at the National Zoological Park, where he had worked briefly in his late teens while on break from the Staunton Military Academy. In 1922 he returned from a Mulford expedition to Bolivia (Mulford was a pharmaceutical company that eventually became Merck Pharmaceuticals) with 40 cages of animals for the zoo, a gift that his wife later claimed led to his being asked to direct the zoo. Any zoo with pretensions of being an elite institution needed an animal collection rich in unusual specimens. Lions, tigers, bears, and elephants might attract crowds, but a scientifically organized collection of rare birds, mammals, and reptiles signified a zoo's status within the zoological community. Mann's worldly experience and academic credentials perfectly qualified him to be a zoo director. So, in 1925, at age 39, Mann accepted an offer to become director of the National Zoo.[16] This career shift was soon followed by his marriage to a woman who shared his love of adventure and felt at home in his scientific community.

Lucile (Lucy) Quarry was born 1897 in Ann Arbor, Michigan. After graduation from the University of Michigan, she moved to Washington, D.C., where she became an assistant editor for the Bureau of Entomology and was introduced to Mann. Desiring a career in publishing, she relocated to New

York City, but did not forget Mann. When he became director of the National Zoo, she wrote him a congratulatory note, and he reciprocated by visiting her while making preparations for the Chrysler Expedition to Africa (a major collecting trip funded by the Chrysler corporation). After a courtship carried on in speakeasies and nightclubs, they were engaged and she traveled to Washington on the weekends, staying at the Dodge Hotel for women. During these visits Bill entertained her with adventurous stories about collecting expeditions. But Bill's interests were not confined to science. He loved opera, could recite Rudyard Kipling and other poets, was a "great reader," and "liked all sorts of people."[17]

Marriage had to be squeezed into a busy schedule. Bill left on the Chrysler expedition for nine months and brought back 1,600 new animals for the zoo. Then he had to convince Congress to appropriate more money to care for the new animals. His old boss and good friend, Leland Howard, asked him when he planned to get married, to which Mann replied, "Oh, as soon as I find an odd moment." To avoid the media attention to their wedding, brought on by Bill's growing fame, the couple had a quiet ceremony on a Saturday morning in October with only Lucile's mother and one of Bill's friends in attendance. For their honeymoon they went to the Philadelphia and New York zoos. Lucile, who had left the Bureau of Entomology because she "wanted something more exciting than bugs," was now married to a man who loved them.[18]

After the honeymoon they moved into an apartment that Bill bought on Adams Mill Road, near one of the zoo entrances. Bill had a five-minute walk to work and frequently brought animals and people home to their apartment. Lucile said that she lived "on the edge of the Zoo" in both the literal and metaphorical sense. Like other wives of zoo directors and animal keepers, Lucile often raised baby animals in their apartment. Susan the lion cub was born at the zoo on July 8, 1930, but her mother refused to nurse her. Lucile took on the job of bottle raising the cub and house training her. Lucile described raising the cub as "carefully and methodically as a human baby."[19] Her daily care included a bath, after which she "had her eyes washed with boric acid; her food was measured and warmed," and she was given "cod-liver oil to prevent rickets." She was "taken out for occasional sun baths," and she "had as much petting and affection as could possibly have been good for her." Within a few months she grew "into a real, honest-to-goodness lion" who would "wrinkle up her face, lay back her ears, snarl and take a piece" out of Lucile's arm if she tried to "dissuade her from tearing strips of leather off [their] favorite chair," or "unhook her claws from a treasured bedspread, or if [she] went to rescue a nervous guest." Over time the Mann home harbored a "kinkajour, a potto, a wart-hog, snakes, turtles, lovebirds, lizards, and several monkeys." Lucile recognized that she was now "wedded to science."[20]

Bill and Lucy were in fact wedded to science, and their passion for the natural world led them to find common cause with like-minded people in an array of clubs and societies. After becoming director of the zoo, Bill's interests broadened considerably, and he became involved in the Vivarium Society, the American Society of Ichthyologists and Herpetologists, the Baird Ornithological Club, the Washington Biologists' Field Club, and others.[21] Through these organizations Mann met other zoo directors and curators, and they addressed issues that were becoming increasingly important to their vision of a zoo man's proper interests.

In 1925, Bill and a circle of friends established the Vivarium Society, which for the next five years met regularly to educate aspiring naturalists (mainly "youngsters who liked to keep cold-blooded pets") about reptiles and the natural world. As president, Mann gave generously of his time, taking young members under his wing to show them around the zoo and mentor them. For five years the Society's 50 or so members met once a month at Mann's zoo office or in a room in the Smithsonian arranged by Doris Mable Cochran, the Society's secretary and an assistant curator for reptiles and amphibians at the Smithsonian.[22] The members included prominent scientists such as H. S. Barber of the United States National Museum, Vernon Bailey from the Biological Society, and W. R. Walton at the Bureau of Entomology. Each meeting began with "a special exhibition of natural history specimens," followed by a short talk or "formal papers."[23]

The Society strengthened social ties between established scientists from various fields and aspiring young naturalists. Their first field trip on July 14, 1925, for example, was as much about socializing as it was about zoology. They went to Plummer's Island where they collected "natural history specimens," including "some mollusks, a young catfish and Daphnia" for Mann's aquarium. Cochran noted that the group saw a "garter snake, *Thamnophis sirtalis*, some green frogs, *Rana clamitans*, leopard frogs, *Rana pipiens* and a painted turtle, *Chrysemys picta*." After the group spent some pleasant time on the island, they went back to Mr. Barber's house for some "cheese-and-lettuce sandwiches and dill pickles" served by Bill Mann and Barber.[24]

For Mann, who was now in charge of a major animal collection, the lectures and activities were surely educational. Dr. Cora Reeves, recently returned from Gingling College in Nanking, China, described the animals she encountered there, illustrating her talk with reptile and bird skins loaned by Cochran.[25] On another evening, biologist Vernon Bailey spoke about an anticipated 300,000 acre government wildlife refuge on the Mississippi riverbanks from southeastern Minnesota to southeastern Illinois. Bailey said that "desirable" animals such as the woodcocks and wood ducks would come back once their habitat was protected.[26] Such lectures addressed conservation issues that would later preoccupy key members of the zoological profession, including Mann.

Mann, who hoped to build a reptile house at the zoo, also joined the American Society of Ichthyologists and Herpetologists, a professional association of zoo directors, government scientists, and college professors. Graham Netting, the secretary of the "Icks and Herps," as Mann referred to the society, was the Pittsburgh Zoo director. In addition to their professional business (such as preventing the U.S. Postal Service from disallowing the shipment of live snakes), the Herpetologists also debunked myths and falsehoods about the natural world. One such target was *Ubangi*, supposedly a film record of a Belgian "expedition into the Congo jungles" from 1925 until 1929. An investigation by the National Better Business Bureau and societies like the Herpetologists, however, revealed that no such expedition had ever taken place. Moreover, the film's zoological fallacies outraged Herpetologist Society members. *Ubangi* was "replete with unnatural history," including "scenes showing a boa constrictor, a tarantula and alligators," none of which actually lived in Africa. The alligators were "misdescribed [sic] as crocodiles and a mygale spider was incorrectly named a baby tarantula."[27] In an era when film was giving an eager public its first views of exotic animals, scientists and zoo professionals were organizing to protect and distinguish truth from fiction.

Mann was also an active member of the American Society of Mammalogists. He knew some members of this society through his professional work (e.g. the Smithsonian's Alex Wetmore and the National Geographic's Gilbert Grosvenor), and many of his fellow herpetologists belonged to both societies. The mammal society extended his network to include a range of other scientists and conservationists working to eliminate steel traps and protect wilderness. When the society held its annual meeting in D.C., Mann contributed money toward the meeting expenses at the highest level ($15), and he also hosted a zoo tour and dinner for the members.[28]

At the same time that Mann was getting to know the herpetologists and mammalogists, he was developing relationships with American and European zoo directors and curators. Not surprisingly, there was some crossover between these two groups. Once he assumed the directorship of the National Zoo, he was in touch with his fellow zoo professionals "all the time." Mann's timing was good. In 1924 a small group of men from the nation's leading zoos had gotten together to form the American Association of Zoological Parks and Aquariums (AAZPA).[29] With his gregarious personality, boundless professional enthusiasm, and first-rate networking skills, Mann soon became an important figure in this nascent organization. Thus, Mann had taken over the

National Zoo just as the men who ran zoos began to think of themselves as professionals with a distinct identity and interests. Mann was one of them.

Creating a Professional Zoo Staff

In the early 1920s, a core of younger, ambitious, and — to a large extent — scientifically-oriented zoo directors and curators took part in the creation of a separate zoological association within the American Institute of Park Executives (AIPE), the national organization to which most zoos belonged. An initial meeting 32 zoo representatives in St. Louis, Missouri, resulted in the idea of a "National Zoological Association." At their meeting, these representatives shared information about their respective institutions and vowed to assist each other in developing America's zoos.[30] By 1924 they had formally organized as the American Association of Zoological Parks and Aquariums (AAZPA). Annual meetings and articles in *Parks and Recreation* (the AIPE journal) followed, and in 1932 the association published its first resource book for the profession, *Zoological Parks and Aquariums: An Annual Assemblage of Information and Facts by the American Association of Zoological Parks and Aquariums.* The AAZPA chairman, George Vierheller, explained that this edited volume filled an "extremely urgent" need: helping planners understand how best to design, finance, construct, and manage a new zoo. As Vieheller stressed, the demand among American communities for zoos was very high, the supply of zoological "authorities" very low.[31] Those authorities — who included the AAZPA Vice-Chairman William Mann and secretary Roger Conant — used the book to stress the necessity of staffing every zoo with skilled employees. The proper zoo would "exhibit animals under only the most favorable and logical conditions," and its director and staff would have the "forethought and expert judgment" to "promote, establish, and administer such exhibits."[32] But, the dean of zoo directors, William T. Hornaday, asked rhetorically, "Where are trained zoological garden zoologists to be found?" His answer was that when a "real demand for such men" is created, "many desirable young men will specially fit themselves to enter" the zoo profession.[33] The book's publication was timely, and as the New Deal dramatically expanded the number and size of American zoos, directors did indeed seek methods of securing staff members who had specialized knowledge about animals.

Most American zoos in the 1920s, especially smaller ones, were run or staffed by city employees without "expert knowledge" of animals. The typical zoo was located within an existing city park, and its management was thus under a park board. The zoo director might be a park superintendant, and the keepers were often park employees assigned permanently or on a part-time basis to feeding animals and cleaning cages. Hornaday lamented the fact that so many cities began by acquiring animals before "securing expert advice and skilled assistance" to run the zoo. In fact, cities built their zoos on the cheap, neglecting to raise money for "expert knowledge" on the mistaken belief that none was necessary. Hornaday saw indifference to "*expert service*" as the "greatest difficulty to be surmounted" if one wanted to have a well-run zoo. Even the "assistant cleaner" should be properly trained by an expert; failing to do so could be "dangerous, and sometimes fatal." Yet, a "strict observance" of civil service rules could conspire against a scientifically managed zoo by forcing a director to accept unqualified employees. Thus, Hornaday insisted that a conscientious civil service board must give the zoo director "complete freedom in the selection and discharge of the employes [sic] for whose acts he is held responsible."[34] Across the nation, zoo directors did lobby their park boards for more control, and as the New Deal construction of zoos created new employment opportunities, young and science-oriented men and women joined zoo staffs, giving them an increasingly professional orientation.

Hornaday made his case for a professional zoo staff in general terms, but to see how these arguments were fleshed out before city officials, we can examine the case of Milwaukee Zoo director Ernest Untermann (1864–1956), who was perhaps the only Socialist zoo director in American history. He was born in Germany and studied geology and paleontology at the University of Berlin. His interest in nature led him around the world to Africa and South America, where he studied animal evolution and collected animals and plants for scientific supply houses. He eventually immigrated to the United States, becoming a citizen in 1893, settling in Milwaukee, Wisconsin, engaging in Socialist politics, studying art, editing a Socialist newspaper, and writing several books. His scientific credentials and commitment to public education landed him the position as director of the Washington Park Zoo in Milwaukee, Wisconsin, in 1935. He was 71 years old. Here he spent five years, finding satisfaction in the animal and educational aspects of his job, but frustration with the personnel management side.[35]

Untermann wanted to professionalize the job of zookeeper, and so he waged an ongoing battle against a budget conscious Park Commission that sought to reduce wages for all park employees. Taking over in the depths of the Depression, Untermann soon faced a Milwaukee Park Commission that proposed to save money by turning all salaried zoo keeper jobs into hourly positions. Two keepers in particular had worked their way up to the point at which civil service rules stated that they should be paid $155 per month salary. The Board wanted to rescind this promise and increase the number of jobs that were never allowed to be salaried. Untermann resisted this effort because of both his sense of economic justice and his understanding of the true requirements of the keepers' jobs. Taking the same position that Hornaday had articulated several years earlier, Untermann argued that the Park Commission underestimated the professionalism of keepers, potentially endangering zoo workers and guests by increasing the number of unskilled hourly workers. Like Hornaday, Untermann insisted that a wise Park Commission would defer all personnel decisions to the zoo director, who alone understood the importance of hiring good workers who could be counted on to appear at a moment's notice. To combat the commission's perception that zoo keeper jobs were effectively menial, Untermann wrote an analysis of how keepers' duties and responsibilities differed from those of other Park Service employees, and he appeared before a joint meeting of the Milwaukee Civil Service and Park Commission on July 5, 1938, to argue for keeping keeper jobs salaried and preventing the job from reclassification into the category of unskilled laborer.

Untermann argued that despite popular misconceptions, zookeepers were skilled workers. He noted that the zoo's central mission made it different from all other Parks Service jobs involving animals. Where other such jobs might involve working with domestic animals, the zoo's central mission required that keepers work with *wild* ones. This distinction was crucial because the wild animals demanded significantly more expertise. Wild animals were "dangerous, even behind bars," so one could not handle them "carelessly" as one might manage domestic stock. Keepers could never take their eyes off of the animals as one could safely do with, say, many horses or cows. Even wild animals that became used to their keepers after years of care could "suddenly turn on them in a fit of rage, maim or kill them." Expertise in handling the animals was life-saving because a "single small mistake [might] cost a human life, injure or kill an animal worth thousands of dollars, and damage valuable property." An "elephant, rhino or hippo on a rampage" could "wreck a whole zoo in a few minutes."[36] Given the costly consequences of entrusting animals to low-paid, untrained workers, the present keepers' salaries seemed modest.

Untermann urged the commission members to consider some of the keepers' actual duties if they doubted their professionalism. They might consider, for example, asking an unskilled man to "go into a Bear cage and pry loose the paw of the animal caught in the bars or in [a] crack in the floor." Or placing a replacement laborer in the position of entering an elephant cage to "pull a tack out of the foot of an enraged animal." They might also imagine the expertise needed to put a hand "into the mouth of the Hippo when it has tooth trouble or a wad of gum in its throat." He invited the commission members to help the keepers dehorn the bucks during deer mating season. He warned, however, not to expect him to pay the bill of the doctor or undertaker after they were carried out on a "stretcher, a cripple or a corpse."

He rhetorically wondered aloud about what the unskilled men would do when a bear "squeezed out of [its] cage and ran into an alley on 42nd Street?" Or "if the Tiger fell out of its cage in the Lion Hall." (Both incidents had actually happened at the zoo during visiting hours.) Untermann knew the answer: the replacement workers would be "blue blurs on Vliet street and cover the distance to the safety of the next house in nothing flat."[37]

These kinds of tasks, and many others that were similarly dangerous, were part of the "daily routine at the Zoo" and were done "without fuss and serious injury to men, animals and properties, because our Zoo Keepers are men of skill and expertise." Thus, $155 was not too much to pay for "men who have the courage and experience to handle wild animals, to feed them right, to take care of their babies, and to nurse them when sick." He pointed out that it took approximately two years to train a new keeper so that he could be sufficiently trusted by himself around animals. He noted that personally, he did not care to "accept the responsibility for what would happen" to these unskilled replacement laborers and proposed that the county would "not enjoy paying large sums for injuries to men and properties." Any cost savings in hourly workers would be lost at the "bung hole of accidents and fatalities."[38]

Unwisely, frugal zoos had experienced exactly these problems when they hired unskilled workers — men Untermann referred to derisively as gypsies and circus roustabouts. The San Francisco Zoo hired a "shiftless circus roustabout" in 1938 who was soon killed by an elephant under his charge. The keeper evidently had tried to take a bull elephant in heat away from some cows, whereupon the bull "spitted him on his tusks and trampled him to death." The bull elephant, "worth thousands of dollars," was put down. In an unnamed eastern zoo a keeper had walked too close to a gorilla's cage and had his arm crushed.[39]

In contrast, Washington Park Zoo visitors typically praised its safety and cleanliness. Untermann, echoing the AAZPA's official line, argued that this was because of his emphasis on professionalism within his staff. His keepers were men of "skill and experience," nearly "all of who [sic] are long-time residents of Milwaukee with families." The keepers "know what to expect from their charges ... understand their animals ... know their feeding habits ... [and] keep a level head in emergencies." If they were replaced, the zoo would "become filthy, the animals ... neglected and unsightly." In addition, the death rate would increase. He noted that on average American zoos during the 1930s lost between 15 to 25 percent of their animals every year, but his zoo "rarely reached that average."[40]

Salaried, well trained workers, moreover, maintained the health of the animals at great financial sacrifice because they were not paid over time. They typically put in "more than 2,000 hours per year without extra compensation." They, and the director, also often had to forgo vacations, Sundays, and holidays, because animal welfare was not on a set schedule. If the animals were sick or there was some kind emergency, the men had to stay on the job. Untermann himself had never had a vacation.[41]

Fundamentally, Untermann and his fellow AAZPA members did not believe that civil service rules should apply to zoo employees. As Hornaday wrote, "a strict observance of the rules of a civil service commission is almost certain to result in the insertion of a large percentage of round pegs in square holes." Instead, a zoo director should be given "complete freedom" to hire, fire, and supervise his employees, and the "more a civil service commission heeds his judgment, the better" for everyone. Untermann agreed wholeheartedly. He chafed at the commissioners' desire to tell him what to do on a day-to-day basis. "If we have to ignore the office rules at times, don't write me a letter telling me we can't do this or that," he told them. "I have to see that they don't go hungry, don't break out and run amuck." If he followed the "letter of the rules all of the way through," his men would soon protest because the rules were made by people who did not understand zoo life. Civil service rules simply had to be ignored when a keeper took a few days off in succession after working overtime or failed to take his vacation when scheduled. The care of wild animals was a professional occupation that did not conform to any predetermined schedule.[42]

Although Untermann's battle with the Park Commission focused on job classification status of lower-level employees, his quest to professionalize the zoo was much broader. Like Hornaday and other AAZPA leaders, Untermann sought to expand the zoo's public mission and thereby elevate its status. For science-oriented directors, a zoo needed to do more than simply provide citizens with recreation. It had a responsibility to educate. A carefully designed and ably staffed zoo could use a variety of methods to transmit important knowledge about the natural world. And, as the Depression wore on, directors made creative use of New Deal programs to support this educational mission.

Scientific Education at the Zoo

The American Association of Zoological Parks and Aquariums opined in its 1932 guide to the profession that a responsible zoo should advocate the "dissemination of facts instead of theories."[43] The public held many misconceptions about animals which needed to be corrected. Fairfield Osborn, president of the New York Zoological Society, argued in 1941 that his zoo's scientific orientation was in the best interests of the average zoo guest, even if that visitor might not know it. Describing the Bronx Zoo, he said its development was:

> ... an affirmation of the fact that the Society, through its scientific work, through its educational activities, is providing the public not only with things in which people find enjoyment or recreation — but with influences that must become more and more a part of the people's conceptions of life as a whole. Man, to his own distress, is obscuring many of the truths which affect all life — including his own. Expressed in its broadest terms, the Zoological Society exists in order to tell people the story and meaning of life on this earth as expressed through the myriad and varied forms of living creatures. It is a symphony of vast and powerful undertones — millenniums of time, evolutionary changes, adaptations to environment. It is at the same time, a symphony of overtones — of beauty of strangeness, of gayety — even of humor. Some day, if our plans and visions may be realized, the words 'zoo' and 'aquarium' will attain a broader significance, one which to-day we can be but dimly aware of.
>
> We are proceeding on the principle that while the present vast preparations of the nation to cope with world conditions are paramount, it is, at the same time, necessary — in a sense more necessary than in normal times — to carry forward those activities which bring recreation and mental enjoyment to the public. Further, our organization is an interpreter of nature, and who is to say that the troubles civilization is cursed with to-day do not arise, in large part, from a lack of comprehension of nature's laws?[44]

Speaking here to the Society's privileged members at the Waldorf-Astoria Hotel, Osborn self-consciously put himself, his audience, and his fellow zoo professionals above the common public. Unlike many politicians, he spoke not of what the people wanted, but of what they "needed," even if they did not know it yet. He and the members of his intellectual class had an obligation to lead the public to enlightenment.

Osborn's sense of zoos' proper mission was widely shared within his community. At the 1930 AAZPA annual meeting, Roger Conant, the Toledo Zoo's reptile curator and educational director, lectured his peers on "The Educational Duty of the Zoological Park." Just as Osborn would do a decade later, Conant drew a sharp distinction between zoological experts and "the average person," whose capacity for learning from books was limited. Instead, these average people could gain far more "by studying and observing wild life alive, even though but for a short time." In addition to labeling exhibits to "teach a true understanding of zoological classification and comparison," Conant recommended that zoological parks nurture education by giving visitors access to trained animal shows, guided tours, public feedings, short classes on "popular zoology," and informative zoo publications. Cultivating strong relationships with local schools was particularly important. Significantly, Conant insisted that zoos had a primary *duty* to provide such education.[45] In various ways, we can see three directors embracing this duty during the 1930s: Ernest Untermann in Milwaukee, Carol Stryker in Staten Island, and Belle Benchley in San Diego.

Untermann nicely explained zoos' educational responsibility in a 1938 lecture on "Modern Zoo Problems" given at a Milwaukee meeting of the AAZPA.[46] Speaking of zoos in general, Untermann believed that over the past 40 years the "character of the zoo [had] changed from a mere menagerie of show animals to a center of scientific research and a link in the educational system." The zoo was expected to "aid in the study of biology and paleontology, an experiment station for the study of animal and human psychology, a character builder, and a refuge for the conservation of valuable animals threatened with extinction." Some of the animals were still trained to perform, it was true, but that was to "exercise their own intelligence and for their relief from the monotony of cage life." Zoos should offer students and scientists "an opportunity to observe at close range those animals which otherwise could not be studied alive, except by the very few fortunate individuals who can make trips of exploration."[47]

Insufficient budgets and public disinterest, however, could thwart even the best educational plans. Untermann pointed to director Belle Benchley and the San Diego Zoo as an object lesson. A pioneer in the effort to emphasize the educational mission of zoos, the San Diego Zoo ambitiously developed "15 informal two-hour periods of lecture and demonstration covering the history, administration and services of zoological gardens." Lecture material covered a review of the world's famous zoos, how animals were shipped, evolution, diseases of animals, how animal welfare was maintained, a discussion of natural and artificial selection, housing, routine daily care, training, and the study of animal psychology. The program included guided trips throughout the zoo and visual aids such as slides, movies, and x-rays. Initially the San Diego Zoo offered this program to all zoo patrons, but budget cuts brought on by the Depression, coupled with what Untermann called the "ignorance of the public concerning the labors and services of a zoological garden staff" forced the zoo to redirect its educational services at a captive audience: school children.[48]

Following San Diego's lead, Untermann instituted a scientific education program at the Washington Park Zoo, where he too found little public demand for a rigorous curriculum. Scientific education required discipline to achieve worthwhile results, but few visitors were willing to submit to assignments and accountability imposed by teachers. So, Untermann pinned his hopes on younger visitors, classes of schoolchildren who could visit the zoo with their teachers. With advance planning and coordination, he could provide visiting students with lectures and guided tours that dovetailed with their classroom curriculum. If the teachers were discussing monkeys or lions in the classroom, for example, they could bring the students to the zoo and meet trained guides in front of the cages for real-life instruction. "Textbook study is dry," and in an era of limited nature films (and none in color), only zoos offered students the opportunity to watch living animals.[49] For guides, Untermann used young men and women from the New Deal funded National Youth Administration; after formal study on particular scientific questions about animals, these young guides met with the school children. In this way, two generations of young Americans were developing their scientific knowledge,

Staten Island Zoo Director Carol Stryker (left) with Carlton Beil, Staten Island Museum curator, and Charles Pearson, Jr., a zoo bird keeper, performing a winter bird count for the National Association of Audubon Societies (1942). This was one of several conservation-related activities in which the zoo officially participated (Staten Island Zoological Society).

with the older group receiving the training that could equip them for future positions in the zoo.

While Untermann used New Deal money to graft educational services onto an existing zoo, planners in Staten Island built an educationally-focused zoo from the ground up. The Staten Island Zoo was the third of New York City Park Commissioner Robert Moses's New Deal zoo projects, announced along with the others in February of 1934.[50] Of the three zoos, Staten Island's was the only one with an educational rather than recreational mission. This difference stemmed mainly from the fact that the zoo's purpose and design were guided by a private non-profit organization — the Staten Island Zoological Society (SIZS) — that intended to manage the zoo once it was built. Modeled on the very successful New York Zoological Society, the SIZS was led by men who were both community leaders and avid naturalists with scientific training. The president, Dr. James P. Chapin, was a staff member of the American Museum of Natural History, an explorer, and a renowned expert on African birds. Carol Stryker, the zoo's director, came from the Staten Island Institute of Arts and Sciences and was an enthusiastic natural history educator and organizer. These men consciously intended the zoo to be more than just a recreational destination. Scorning the typical zoo as offering "little beyond the thrills of a vast animal pageant," Stryker and SIZS vice-president Harold J. O'Connell promised that the Staten Island Zoo would serve the "increasingly serious interest in natural history on the part of the public." In practical terms, this mission meant exhibits that featured reptiles and smaller mammals and birds, many indigenous to North America, which could be viewed and studied in an intimate setting rich with science-based interpretative materials. Moreover, the new building would divide its space between exhibits and rooms for study, research, and naturalist club meetings. The SIZS leaders planned to use the zoo as a community center for teachers, students, and other members of the public who had a desire to learn more about the natural world. Where the typical zoo provided only a short recreational experience, the Staten Island Zoo hoped to "place at the disposal of the public facilities of lasting interest for the use of longer periods of leisure."[51] Here then was a different model of "leisure" than that championed previously by smaller zoos. As in Milwaukee, San Diego, and elsewhere, this model — which stressed the development of scientific knowledge — received crucial support from New Deal relief programs, particularly the National Youth Administration.

Science, *research*, and *education* were the operative words in describing the new direction zoos like Staten Island's were taking in the 1930s. To conduct animal research and expand scientific knowledge, zoos required new kinds of spaces: laboratories, hospitals, and classrooms. The San Diego Zoo opened its Biological Research Institute to perform *post-mortem* examinations of zoo and wild animals, investigate nutritional requirements, and study parasitic diseases. The zoo housed the Institute in its existing hospital building and organized it along professional lines, filling a research committee with Ph.D.s from local universities and offering two annual research fellowships.[52]

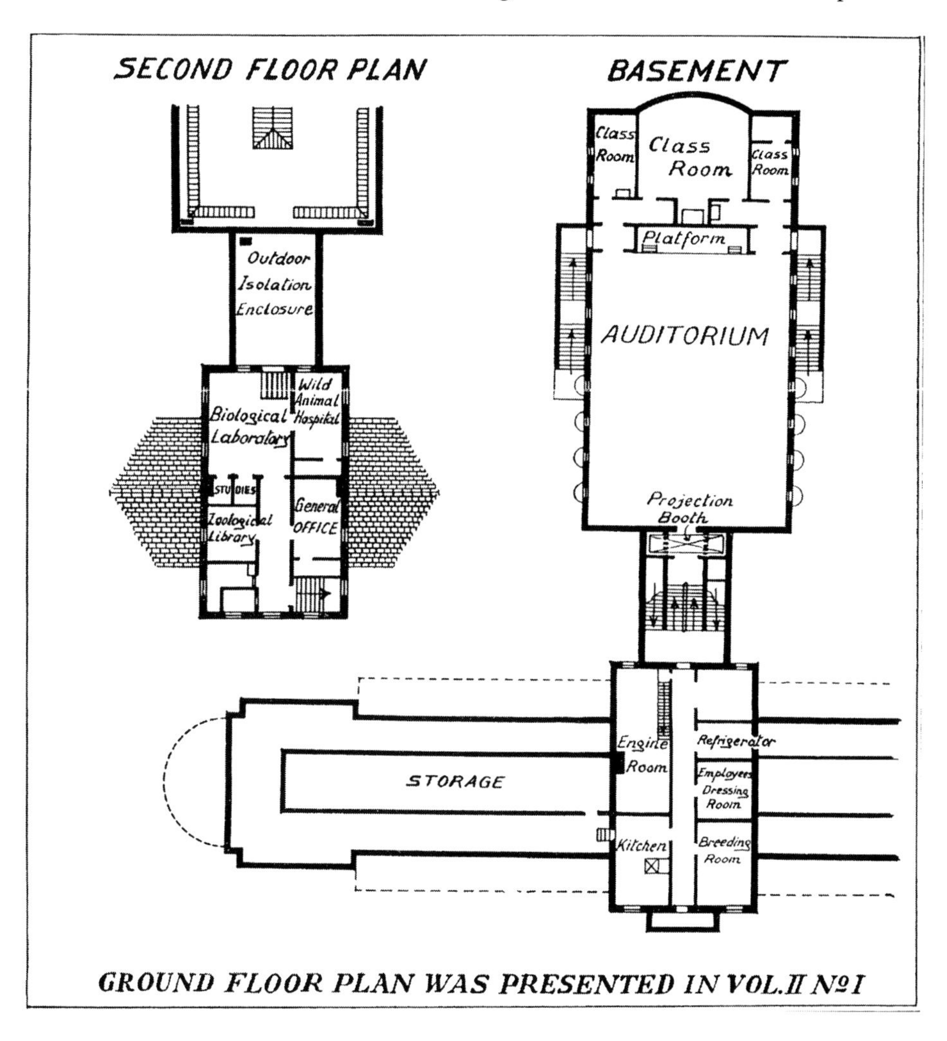

Second floor and basement plan for the Staten Island Zoo (1936). The basement auditorium and classrooms served the zoo's educational mission, while the second floor library, laboratory, and hospital served its science mission. Both kinds of activities were promoted by the Association of American Zoological Parks and Aquariums and we made possible by New Deal construction and employment programs (Staten Island Zoological Society).

During the same decade, the Philadelphia Zoo, which had housed the Penrose Research Laboratory since 1901, turned its attention to animal nutrition, and the New York Zoological Society formulated plans to become the nation's foremost researcher of animal diseases.[53] The Staten Island Zoo joined the ranks of such institutions with its own laboratory where it too undertook the study of animal nutrition and disease.

Shaped by SIZS's science orientation, the zoo's design process differed sharply from that utilized in the city's two other New Deal zoos. In approaching the Central and Prospect Park projects, Moses favored speed over all else. His design teams worked quickly, producing aesthetically pleasing zoos that, although new, were not particularly innovative in terms of cutting-edge technology or exhibition design. By contrast, the Staten Island design team spent months researching other zoos, ensuring that its zoo would "represent the last word in modern zoological construction development" and serve as a "model" for future zoos. The zoo's reptile wing required particular design attention, due to the special temperature and safety requirements of its inhabitants. The inclusion of designated spaces for a laboratory, hospital, auditorium, and library indicated that the design would serve the interest of animals and science, not just of recreation. Even the role of WPA artists reflected a fundamentally different design philosophy. In the Prospect and Central Park zoos, WPA artists enhanced the architecture, adding bas reliefs, sculptures, weathervanes, and other elements all intended to provide visitors with visual stimulation in addition to the live animals. On Staten Island, the WPA artists focused on the exhibition spaces, painting carefully researched murals on the walls of each cage to help visitors understand the animals' natural habitats.[54]

The planned Staten Island Zoo differed not just in how it intended to serve the public interest but also in how its funding and management design required intense outreach for public support. Because the SIZS would manage the zoo, city leaders had to seek state legislation authorizing the city to fund annual maintenance; the resulting bill, which made the mayor and park commissioner ex-officio members of the SIZS board, was signed by governor Herbert H. Lehman in May of 1935, and a week later the Governor reiterated his support for this kind of public-private partnership by donating a "fine beaver" to the SIZS.[55] Even before construction on the zoo had begun, the SIZS was recruiting members, with an Associate Membership costing $3 and a Fellow Membership costing $50. Although CWA funds would be used to build the new zoo and the city would pay for the lights, heating, and staff salaries, the SIZS would be responsible for purchasing collections and supporting ongoing lecture, educational, and research programs. Thus, in its membership campaign, the SIZS appealed to "public spirited and forward-looking citizens [to] recognize the value to the community of this undertaking and accept their full share of responsibility by joining" the society. This appeal for public support was relatively unusual in an era when most public zoos offered free admission, but it was highly recommended by the AAZPA, which pointed out that educational programming and publications — for which "city functioning bodies rarely have money" — could be sponsored by a zoological society.[56]

The SIZS's efforts to generate public support were successful. Over 500 local residents attended a garden party in 1934 at the site of the future zoo to hear about the design, and many more attended lectures and other events over the next few years.[57] To raise money for maintenance, the society hosted a charity "Jungle Ball" in November of 1935. Over 2,000 guests enjoyed post–Prohibition cocktails such as "Snake Bite" (Angostura bitters, grenadine and rye), sampled bear steaks and other wild game, danced the "Monkey Hop," viewed WPA artist Olive Earle's watercolor paintings, and scared themselves by peering into a venomous snake pit.[58] Citizens also donated numerous animals to the SIZS, even before any buildings existed to house them.

Although the zoo's physical construction progressed at a snail's pace, relief funds allowed a "virtual" zoo to operate in the absence of permanent buildings. Even while work on the zoo proceeded in fits and starts, Carol Stryker kept CWA sponsored zoo employees (some assigned apparently to the Staten Island Institute of Arts and Sciences) busy caring for a growing collection of reptiles, small mammals, and birds that were housed in a temporary barn and other locations around the city.[59] Throughout the several year construction period, Staten Island newspapers identified these men as "zoo staff." Frank Dickson, an animal keeper, was the most prominent, appearing frequently in stories about the zoo's growing snake collection.[60] When the zoo finally opened in June 1936, Dickson became head keeper, and at least four other former relief workers assumed full-time positions.[61]

Soon after opening, a minor crisis with the zoo's operating budget foreshadowed the kinds of problems that most public zoos would not face until decades later. Designed by the SIZS as a "modern" facility with an education and animal welfare mission, the Staten Island Zoo required more than just park employees to feed the animals. Its educational facilities needed instructional equipment and scientifically trained curators to direct its programs. Its on-site animal hospital — unusual for the era — could function only with a budget for laboratory equipment and supplies. Unfortunately, in an era of limited revenues, the city's Board of Estimate slashed the promised funding for these purposes from $39,000 (for seven months) to $23,100 — only a third of the original budget request.[62]

Luckily, federal relief programs, including the National Youth Administration, could be tapped to supply educational and other staff that enabled zoos to fulfill their scientific and educational missions. In San Diego, the zoo's Biological Research Institute enjoyed the services of several San Diego State students compensated by the NYA to carry out "technical and statistical investigations under the direct supervision" of the Institute's director.[63] On Staten Island, the zoo opened with a WPA staff that numbered nearly 30.[64] The significance of this enlarged staff was that it allowed the zoo to fulfill its education mission; without it, noted O'Connell, the zoo was "only slightly better than a menagerie."[65] In 1937, the SIZS reported that WPA workers were providing vital services in the areas of osteology (mounting snake skeletons), education, accounting, laboratory technicians, stenography, sculpture, painting of cage labels, and other art work. Without these supplemental employees, the SIZS concluded, the zoo would have accomplished much less of its planned work. Indeed, given the fact that WPA workers wrote news releases, compiled data on specimens, photographed the zoo's inventory and activities, prepared specimens for biology classes, and performed administrative support services, one could argue that WPA relief labor *enabled* the Staten Island Zoo to be more than just a menagerie where public employees fed the animals and cleaned their cages. The 11 NYA employees, for example, gave the Staten Island Zoo nearly 500 hours of service a month, and even as late as 1939 the zoo had nine WPA workers assigned to it.[66] This relationship provided good public relations material for the relief agency, as illustrated by a WPA "open house" week in the spring of 1940 at which signs and newspaper publicity reminded visitors of the WPA workers' accomplishments.[67]

The WPA in New York contributed to zoos' educational mission in other ways. Writers for Federal Writers Project compiled four natural history books that drew heavily on zoos for research and illustrations: *Who's Who in the Zoo*, *Reptiles and Amphibians*, *Birds of the World*, and *American Wildlife*. Research for these books was conducted largely at the Bronx and Staten Island zoos, and the photographs were taken of specimens on exhibit at the two zoos.[68] The books sold well, allowing zoologists to reach a wide audience with their expertise.

Thus, forward-minded zoo directors during the New Deal used federal funding and their own belief in science to push the mission of American zoos toward education. While they hoped to educate all patrons who entered their gates, they had the most success with school children, who (to varying degrees) could be forced to take tours and listen to lectures. By establishing good relationships with local school systems, directors hoped to make zoos an arm of the American educational system, conveying scientific knowledge of exotic animals to young members of the middle and working classes who could not afford to travel to foreign countries to see the animals in their natural habitats and whose parents were not versed in the scientific facts of every animal on display at their institutions. And schools and teachers responded. In Seattle, for example, the Woodland Park Zoo director's monthly reports typically offered a long list of teachers who had brought classes to study everything from feeding behaviors to natural habitats; the director enthused that "to be able to study the lessons along with the live specimens makes the schoolwork more impressive as well as more lasting."[69] The *idea* that zoos were educational had been voiced for decades, but it was in the 1930s that the idea was put into wide practice.

Conservation and Natural Selection

Young Americans were a natural and primary constituency for zoo's educational efforts, but adults held political power and could not be ignored. Those zoo professionals with a scientific orientation recognized that both animals and science itself were threatened by public ignorance. On the one hand, Americans who did not understand and respect the role of wild animals in the ecosystem were likely to exterminate species and destroy habitats. Thus, zoo directors and curators found themselves increasingly serving as authorities and even advocates on conservation measures. On the other hand, the zoological sciences rested on the theory of natural selection, which was under attack from some quarters of society. So, zoo professionals also found themselves thrust into the role of explaining and defending this controversial theory. In both situations, we are reminded that the men and women who ran the zoo belonged to an elite class of Americans. As Hornaday wrote, the "far-seeing and broad-minded" members of the zoological community would often have to struggle against the "proletariat" and the "stupid people" who could not recognize the "public benefits" of a zoo; by implication, those same people would resist conservation and science unless led by "men of public spirit who appreciate wild life."[70] For Hornaday, the men who ran zoos were public-minded leaders.

Members of the New York Zoological Society had always given a lot of attention to conservation issues, but in the 1920s and 1930s, a broader range of the national zoo community began to speak out against the destruction of wildlife and habitats. The AAZPA explicitly advocated the "conservation of wildlife," counting efforts in this area as among "the great influence for good" that could be "given by the right kinds of zoos."[71] Conservation organizations reached out to these kinds of zoos, finding natural allies among their leaders. In Seattle, for example, the National Association of Wild Life Conservationists and the Audubon Society both donated rare animals to the Woodland Park Zoo, which justified its exhibition of endangered species as "the only means of preserving these specimens" until the "coming generation [can] be taught to preserve rather than destroy."[72] Many directors were quite comfortable with seizing the opportunity to provide conservation education.

Zoo professionals who spent considerable portions of their time collecting and researching animals in the field, saw firsthand the wanton destruction of species by habitat loss and human predation. Conant's encounters with people during his multi-year survey of Ohio reptiles convinced him that most Americans killed all reptiles, harmless and venomous alike, on sight. Newspapers in Tennessee and Missouri reported weekly rattlesnake and water moccasin "kills" during the annual fall and spring migrations of those reptiles to and from their wintering grounds in southern Illinois.[73] Researching Lake Erie's water snakes, Conant gained access to a private island whose owner was elated that Conant's team found only one snake, evidence of "how efficient we've been in getting rid of the snakes here."[74] Carol Stryker and colleagues conducted annual surveys of Staten Island's wildlife, and in an article about the research, reptile curator Carl Kauffeld noted resignedly that of the seven snake species on the island, he expected only three could survive much longer because of human destruction of their habitat. Ironically, New Deal public works projects — dams, swamp drainage, road construction, and so forth — caused both the direct destruction of habitat and hastened its eventual destruction by providing an infrastructure for expanded human development.

In Tulsa, Oklahoma, in 1927, Will O. Doolittle — who had just become the general superintendent of the city's parks — proposed the creation of a zoo that would be infused with his conservation ideals. Greatly influenced as a young man by the example of William T. Hornaday, Doolittle was a passionate naturalist, a man who believed that close contact with nature made people better citizens. A zoo, therefore, was of great public benefit. As Doolittle made his case for a new zoo, two of his central arguments struck a conservationist chord. First, he claimed that a zoo would "aid in preserving many species of native mammals, birds, and reptiles that are approaching extinction"; and second, he argued that a central purpose of the zoo would be "to secure better protection of animal life by educational methods."[75]

Indeed, as the Tulsa Zoo took shape under Doolittle's direction, it did champion conservation causes. An onsite natural history museum provided education, staff conducted conservation education programs, and the zoo exhibited local birds of prey (which were often shot indiscriminately) to point out their value in controlling rodents and insects. More significant, zoo staff took practical and political steps to protect wildlife. Director Hugh Davis, who also became president of the local Audubon Society, worked with Doolittle during the 1930s to protect waterfowl in the Tulsa area. Zoo employees helped Audubon Society members band over 10,000 migratory ducks in 1939 and 1940; others photographed rare Oklahoma birds in a bid to preserve their habitat; and Doolittle and Davis lobbied for laws to protect birds of prey, prevent the importation of wild bird plumage, and create wildlife sanctuaries. And, some of these political efforts were successful.[76]

Doolittle's conservation influence reached beyond the Tulsa area. A longtime member of the American Institute of Park Executives (AIPE), Doolittle played a leadership role in that organization as well as in the AAZPA. He served on the AIPE editorial staff and edited the AAZPA's first guide. Hugh Davis edited the zoological parks section of the AIPE's *Parks and Recreation* journal, and he also was an AAZPA board member and served as its conservation chairman.[77] Together, these two men and their small zoo demonstrated that a conservation ethic was moving beyond the major east coast institutions.

Zoo directors' desire to protect wild animals from extinction was noticed by men like John Holzworth, a New York attorney who worked to preserve American grizzlies from extinction. Before joining the law firm of Delafield, Thorne & Rogers, Holzworth had been employed by the U.S. Biological Survey to lead scientific expeditions into British Columbia and Alaska. This experience convinced him that grizzly bears, among other animals, needed saving. To that end he started a movement to create a wildlife sanctuary on Alaska's Admiralty Island.[78] Thus, in 1932 he was the New York Zoological Society's Chairman for their Alaska Bear Committee, and he represented the Society of Mammalogists in lobbying the U.S. Senate for an Admiralty Island preserve. Holzworth sought the preserve status because of brewing plans to exterminate all of the island's bears (through hunting, trapping, and poison) to make it safe for logging.

Holzworth's bear campaign had the full approval of William Hornaday, former director of the Bronx Zoo. Hornaday likened Holzworth to J.A. McGuire, a naturalist and writer for *Outdoor Life Magazine* who led a decade-long campaign to save the bears. In Hornaday's opinion, Holzworth was "another able champion" of the bear who "modestly steps into the arena to exhibit the splendid personality of the bear."[79] He was certain that most Americans did not want to see the bears exterminated, but noted that "once upon a time," there were "a few malcontents in Alaska [who] set out to exterminate the Big Brown Bears on a basis of about $25 per hide."[80] Hornaday accepted culling animals when their populations exploded, but he was ardently against the idea that any species was a pest deserving of extinction: "as for extermination,— never! And no killings of great game by poison if you please!"[81] Too often, he knew, government agencies had supported such exterminations, and he took aim at American farmers who killed wild animals if they threatened livestock, urging them "not to appraise [the bear's] value in terms of sheep and goats."[82]

As Holzworth tried to build support for his bear preserve plan through radio addresses, he read prepared statements by Hornaday testifying to the need to preserve the bears. Hornaday warned listeners that "as matters stand to-day, the great brown bears of Alaska must immediately have better protection or perish." If the bears were exterminated "it [would] be a permanent shame and disgrace on the people of the United States."[83]

Hornaday was not the only zoo director who offered unflinching moral opinions on conservation issues. In Milwaukee, Ernest Untermann was just as blunt in his condemnation of Americans' callous disregard for wild animals and their shortsighted lust for material gain. Regarding the animal and environmental destruction evident all around him, Untermann noted that civilization had "made the land of this continent very sick." In Americans' "eagerness to get all the money we can in the shortest possible time, we ... have over-plowed, over-grazed, over-lumbered our heritage. We have almost skinned poor Mother Earth alive; she is covered with wounds of our making."[84]

Like Holzworth and Hornaday, Untermann lashed out at hunters, the U.S. government, and business interests. He had no patience for "sportsmen" who posed as "friends of wild animal conservation," but in reality continued to reduce animal populations, assisted by corrupt game wardens and self-serving hunters' societies. Government conservation efforts were typically weakened by a desire to appease and even promote powerful economic interests such as gun and ammunition manufacturers. He lamented that "useful" predators and birds of prey suffered "wholesale destruction" with full government support. He hoped that zoos could become centers "for the dissemination of this understanding, a strong link in a real conservation policy." He knew though, that too many dismissed his "obvious truth" about the threats to America's wild animals as "radical propaganda."[85] As Untermann was well aware, his views — both moral and scientific — were not shared by large numbers of Americans.

If zoo directors had difficulty convincing their fellow citizens of the need to protect animals, they entered even more treacherous terrain when they discussed the theory of natural selection. The nation had recently been through the Scopes trial, which highlighted the significant differences of opinion among Americans about the science of evolution and the extent to which it ought to replace creationism within American public schools.[86] Because zoos displayed the very animals under discussion in these debates, directors were on the front lines when it came to the popular discussion of Darwinian theory. Zoo directors from progressive zoos clearly articulated their belief in evolution, their struggles with patrons who believed in creationism, and their desire to turn zoos into educational laboratories for public schools. They also undertook efforts to educate their visitors about evolution through interviews and their own writing. Through their private thoughts on patrons who believed in creationism, we can see the ideological clash that occurred between elites at zoos and the people they served.

Belle Benchley, the director of the San Diego Zoo and one of the few women who directed a zoological institution during this era, discussed the importance of Darwinian theory in her 1942 memoir. Writing for a popular audience, she described evolution as "a long and arduous road beginning with the first simple cell which, by dividing itself into two living parts, performed the initial miracle of reproduction."[87] As she explained, after emerging from the jungle, "primitive man found himself able to stand erect and walk upon two feet, bearing burdens or his helpless young in his arms."[88] For a potentially skeptical audience, she argued that the zoo was itself evidence of evolution. After man finally "progressed to that degree of civilization which enabled him to become interested in creatures" for "what we call scientific interest ... he went out into the world and brought back to his dwelling places beasts from far away."[89]

At their best, modern zoos could combat the erroneous and unscientific ideas held by many visitors. As Benchley pointed out, explorers who went into the wild often brought back sensational tales of apes, knowing that these would sell better than scientific observation. And, even among scientists, the term *primate* had long been used to designate "the highest of the creatures preceding man."[90] But what zoo visitor would realize that "the word primate has been enlarged in meaning to include man himself"?[91] Many, she hoped. To help those visitors understand their relationship to the non-human primates on display in the zoo, she carefully classified the anthropoid apes, beginning with the gibbons because they were farthest removed from man and ending with gorillas, which were the closest. She noted with some pride that one day a visitor was watching her two gorillas play in a way that looked like rounds of fighting. Benchley recalled one of the men saying, "Who says Darwin was wrong? That's where prize fighters came from."[92] A zoo held the potential to convince Americans of the truth of Darwin's theory, even if Clarence Darrow could not.

As creationists might have expected, at least one proponent of evolution–Ernest Untermann — was also a critic of religious ideas about the natural world. In an article he penned called "Our Friends at the Zoo," Untermann described the long history of ignorance of the natural world. He pointed out that many myths about the natural world came from "second hand information" that was "distorted by religious prejudices."[93] Alluding to the fact that some scientists combined their religion with their work, Untermann stated that "scientists, especially of the 18th and early 19th centuries, gave us a lot of inaccurate or altogether false information about animals" because we could "not expect anybody to give us reliable facts about the distribution of animals, so long as he believes that animals had been created as they are today and never changed."[94] Sadly, until about a century ago, "natural science was still under the influence of state and church authorities." Since then, however, science "emancipated itself from such handicaps and won the right to follow the facts where they lead." Science, he believed, had "become a democracy." Its purpose was to free mankind from fear, "raise the new generations to the status of self reliant race, to conquer nature and to face all facts with a courage born of knowledge." From his perspective such a "democratic science" was crucial to a democratic government.[95]

Like Benchley, Untermann tried to explain science to a popular audience. In one article, for example, he explained to readers how scientists know that animals evolved over time. His background in geology helped him explain how the scientific investigation of geological strata uncovered marine shells on the summits of mountains and the skeletons of dinosaurs in freshwater sandstone, revealing important clues about how the earth had changed over time.[96] Similarly, he was interviewed about popular animal myths, one of which was that hippopotami "sweat blood." He reminded his audience that this story was "started by the Bible and fostered by circus press-agents who wanted to get money out of the public's pocket." The red that appeared in small watery drops was simply an oil that acted as a waterproofing agent for the animal.[97]

The frustration with visitors' ignorance about science, and the gap between the elite zoo directors and their patrons, often came out in zoo directors' private reflections. While the rest of the country was caught up in the Scopes trial, Benchley was calmly espousing Darwinian Theory for a mass audience. Recalling her first years at the zoo, she noted that "almost daily" visitors asked her if her experience with nonhuman primates "taught her that man really did descend from monkeys."[98] She noted that the question was generally asked in a deprecating manner as though the "speaker were either facetiously playing up to a pseudo scientist or giving me credit for far too much intelligence to believe such a ridiculous theory."[99] In anger at her visitors' lack of scientific understanding, she wanted to reply that if Darwin had watched the behavior of some of her visitors, he would have reversed his theory on evolution and concluded that man actually "*descended*" from the apes.[100]

Untermann also despaired at his guests' lack of scientific understanding. Explaining his dullest visitors, Untermann argued that from his perspective "food, sex and brains" had been "a bothersome triangle for the human race from its earliest origins." Despite bigger brains, humans had not learned to control either appetite by self-discipline and had therefore failed to prevent the "breeding of inferior human strains." The "selfdisciplined [sic] types" have always been "outnumbered by the inferior masses," who are "ruled by their appetites for food and sex." As a result, they weighed too much and brought too many children into the world, from Untermann's eugenic perspective. This doomed the "small superior minority" to a hopeless struggle against the "flood of inferior types." He believed that too much sex sapped human brain development and that families with many children were unlikely to invest much in their education. In a more mainstream argument, he proposed that decreasing the number of children tended to increase the amount of money in the family and would likely result in more education. [101]

Zoo professionals were enthusiastic about their institutions' potential to do good. Some, like Untermann, imagined themselves beating back the frontiers of ignorance in a heroic struggle to save the planet's animals and elevate the position of science. Others, such as Benchley, bit their tongues and pinned their hopes on the theory that visitors would learn compassion and understanding simply from the experience of observing real animals. And still others, like Doolittle and Davis, threw their energy into various methods for realizing conservation goals. All, however, were also in the entertainment business, and even as they advocated education, science, and conservation, they also made sure that visitors had a good time.

Conclusion

An orientation toward research, conservation, and education potentially threatened to distance zoos from the public they served. Recognizing this possibility, Robert Moses, ever desirous of pleasing people with his public works, urged the director of the Museum of Natural History to forgo scientific accuracy in favor of language that the average American could understand. Scientists had to be careful not to alienate the average citizen, because as Moses stressed, "you will increasingly need public funds and, therefore, you should not befog your patrons."[102] Still, zoo directors often used the authority of elite and better-educated Americans to gauge the value of their institutions. In Seattle, for example, director Gus Knudson peppered his monthly reports to the Park Board with references to the views of these "better" visitors. In May of 1933, the National Parent-Teacher Association held its annual meeting in Seattle and many members visited the zoo, whose "wonderful educational value" they praised. As Knudson observed, "this class of people should be well versed on this question."[103] One wealthy, world-traveled woman felt that her young son "derive[d] great benefits" from seeing zoo animals.[104] Such affirmation from the community's "better" citizens was fine, but like Moses, most zoo directors gauged success by their institutions' popular appeal. Here again, William Mann served as a model leader, seamlessly combining high-brow scientific credentials with low-brow enthusiasm for popular entertainment.

Mann enjoyed people, not just animals. At home within a circle of the nation's foremost scientists, as well as prominent artists and politicians, he was gregarious, the kind of person who joked that the safety chain on his apartment door was there to make sure that "if a burglar got in he couldn't get out again, because he'd rather have a burglar than nobody in the house." This quality perfectly fit his role as director of the National Zoo, where the overriding goal was to make the public feel welcome. Lucile recalled that if Bill saw a Montana license plate in the zoo parking lot, he would search the grounds until he found the driver and then invite him home for dinner. And Mann was no snob when it came to dining. If Lucile did not have enough in the cupboards to feed the unexpected guests, he sent out for Chinese food. When the visitors were international scientists in town for a conference, he had no qualms about sharing a dinner of canned salmon salad and chocolate cake, however mortified Lucile might be by the odd combination.[105]

Mann also shared the average American's passion for entertainment, never losing his childhood wanderlust for the circus. Even after becoming "Dr." Mann, he remained an enthusiastic fan of all circuses, large and small. He adored them, often organizing his travel and work schedules so that he could attend them. Reportedly, he was so well-known in the circus world that whenever he attended one that had made a stop in Washington, a long line of circus personnel would form to greet him.[106] Lucile discovered that marriage to Bill meant going to the circus every time one was in town — and not just once, but to every performance. Even if the Manns were at a dinner party, Bill would start telling stories about the circus and before long the entire group would troop off to catch the show. Later, he might invite someone from the circus home for lunch or dinner.[107] Lucy liked to joke that on any given day Bill might come home with the president of Harvard University or a circus clown.

To many among the elite, the circus hands and their typical audience were from a different social world, but Mann counted good friends among their ranks. He visited the circus not just because he liked animals but also because he enjoyed socializing with the circus workers. Being a frequent guest and naturally gregarious, he soon became well known within the circus world. He and Lucy would spend hours visiting with performers; owners invited them to stay in their homes; the famous circus clown Emmett Kelly became a lifelong friend. Henry ("Buddy") Ringling North was on a first-name basis with Mann. As Lucy said, circus people might be a bit "standoffish" at first, but once they got to know you, they were wonderful company. Through the circus Mann met other enthusiasts. Dr. L. C. Holland, an ardent circus fan, struck up a conversation with Mann at a Ringling show in Norfolk, Virginia, and later sent him a copy of a long verse tribute to the circus that he had authored.[108]

The Manns' friends and acquaintances were drawn into Bill's circus passion. Tangential references to circuses appeared in letters from friends and colleagues. It is hard to know if Mann's friends took the circus as seriously as he did. When he scheduled Bill Morden (American Museum of Natural History) to give a lecture at the annual Mammal Society meeting on the mistaken assumption that the circus would be in town, he called the mistake "an awful thing" and "lay awake all night wondering whether you were coming to the circus or the Society meeting, or both." Once a year, when the Ringling Bros. show was in town, the Manns would host a circus party for as many as 75 guests. After a catered dinner at the zoo (sometimes prepared by Lucy), the guests would all go the circus, where Bill had reserved an entire block of seats. Typically only a few members of the group were true circus enthusiasts, but the annual party was grand — one year a congressman arranged for a police escort — and it brought Washington's elite together with the "ordinary" Americans whom they served in politics or science.[109]

Mann's democratic taste in friends and his passion for both science and amusement made him particularly well-suited to guide the National Zoo, and made him a model for other zoo directors. He fully shared the general public's desire to be entertained by exotic animals, yet he was also committed to advancing a scientific understanding of the natural world. For Mann and his like-minded colleagues, a zoo could perfectly combine amusement and science on a grand scale, providing a unique platform from which to educate large numbers of citizens. But as the Great Depression approached, Mann and his colleagues faced a significant challenge: how to fund the construction of the wonderful institutions they imagined.

During the roaring twenties, scores of new zoos were established across the United States. There was a popular enthusiasm for zoos, and on paper it appeared that America was becoming zoo rich. In reality, local governments found it difficult to support more than modest menageries, and even Mann's National Zoo was woefully underfunded. Mann feared that the National Zoo would never become the first-class institution that he desired. Ironically, the Great Depression and the arrival of New Deal work relief programs gave Mann the opportunity to go on a building spree. The New Deal quite literally allowed him to build the zoo of his dreams. And this tale of a zoo rescued by the New Deal was repeated across the nation during the 1930s.

THREE

Why Zoos?

Tired of war and politics,
Folks a-throwing mud and bricks?
Tired of rationing and graphs,
Taxes, scarcity, the draft?
Well, it's spring—here is what to do
To relax, go to the zoo.

—Toby LaForge, *Tulsa Tribune*, no date[1]

Why zoos? In the midst of severe unemployment and hardship, not all Americans understood how building new zoos served the public interest, nor were they particularly concerned about the welfare of exotic animals. Soon after Robert Moses announced his intention to renovate the Central Park Zoo, the chairman of the Emergency Committee on School Overcrowding questioned why $411,000 should be spent on "a new monkey house" when new school buildings were a more pressing need.[2] One letter to the editor protested the planned Staten Island Zoo as "ridiculous on the face of it. Human beings are allowed to live under miserable conditions" while "zoologists" seek to house animals in a "handsome" new structure.[3] Some citizens in San Antonio objected to the city's investment in "motels for bears and monkeys."[4] An anti-tax group member described the Detroit Zoo as a place where "idle animals are kept in luxury while Detroit people starve."[5] This group's leader predicted (incorrectly) "that before the dreadful year 1931 was over, the people of Detroit might be eating the Swans and Monkeys."[6] This inaccurate prediction formed the basis for the later belief that Detroit citizens killed and ate their zoo animals. While this never actually happened, these critics articulated an important political question: Why should the government spend tax money on zoos when those funds could instead be used to provide relief for America's unemployed and poor?

Such critics, of course, often harbored deeper convictions that taxpayer financing of relief programs was wasteful in general. Cost overruns on New Deal projects provided specific evidence for this charge. In New York City an Aldermanic Investigating Committee tried, unsuccessfully, to make a scandal of the fact that all three of the city's New Deal zoos ended up costing far more than originally promised. The Central Park Zoo bill went from $411,000 to around the one million dollar mark, with some arguing that the real cost was closer to two million; similar figures were given for the Prospect Park Zoo.[7] The Staten Island Zoo project had begun as a $65,000 building, but was finally completed for around $500,000.[8] As we shall see, Robert Moses too decried the inefficiencies of federally-funded public works projects, but even though he railed against inferior workmen and wished that he could construct on a contract basis, he steadfastly denied that New Deal public works projects were *inherently* wasteful. Moses defended the WPA against one skeptic, pointing out years later that 12 New Deal recreation centers built in New York City were "alive with kids today, and . . . certainly don't represent boondoggling or make out Harry Hopkins to have been a mere social worker."[9] In his view, the city *needed* three new zoos, and their addition was a cost-effective way to provide a vital public service. Like Moses, newspapers and politicians in city after city emphasized the "importance of the zoo for educational and recreational purposes."[10] This view dominated the popular reception of New Deal zoos.

Questions about how to justify taxpayer support had dogged zoos even before the arrival of New Deal programs, and these questions partly explained why so many city governments had given these institutions only small budget appropriations. Zoo advocates, however, had been honing a class-based defense of zoos that gained new urgency and relevance as the effects of the Depression took hold. This line of defense emphasized the zoo as a people's institution. Ordinary citizens derived great pleasure and educational value from looking at animals. The people wanted the zoo, so the democratic system should respect their desires.

The central question that New Deal zoos raised on the issue of class was this: which part of the public mattered? Was it the very needy who were sick and desperately poor? The so-called unable-bodied poor, who were physically unable to support themselves? Or was it the deserving public, generally thought of as male workers, who through no fault of their own were consigned to enforced leisure? The answer that cities came to was much like the general approach to social welfare in America, which ultimately placed more emphasis on helping the deserving poor rather than offering social assistance as a right of citizenship.[11] Zoos should be funded because they served the right part of the public, which in this case comprised children who were poor through no fault of their own and workers who were suddenly unemployed because of the Depression.

Despite the occasional criticism about cost, zoo projects, because they were photogenic and immensely popular, generated useful publicity for the New Deal and its broader public works campaign. To this end, photographs and models of the zoos were utilized to give the public clear visual representations of the physical transformations made possible by federal funding. Only two months into his reign as park commissioner, Moses provided the *New York Times* with seven architectural renderings of the new zoos planned for Central Park, Prospect Park, and Staten Island. Three-dimensional aerial drawings, with their carefully numbered and labeled buildings, impressed upon New Yorkers how significantly their zoo-going experiences would be improved, both quantitatively (the large number of exhibits and animals) and aesthetically (the beautiful new structures and landscaping).[12] When Moses reviewed the accomplishments of his first year on the job, the *New York Times* highlighted all three zoos, even though two had not yet been completed.[13] Describing the CWA's work in New York, the paper pointed out that the state's final report identified the Central Park Zoo as "among the most constructive" and "especially notable" CWA projects in the city.[14] The zoos figured prominently again in June of 1935, when the Municipal Art Society sponsored a "Park Exposition" designed "as a tribute to the work of Commissioner Moses and his staff." Jiggs and Anna, two Central Park Zoo chimpanzees, were scheduled to entertain Exposition guests near a photo display contrasting the old and new Central Park Zoos as well as the Prospect Park and Staten Island zoos, which were still under construction.[15] When the *New York Times* ran a laudatory retrospective on Moses in 1938, the lead photograph was an aerial view of the Prospect Park Zoo.[16] WPA artists assisted in these booster activities, crafting a scale model of the Staten Island Zoo for events such as a Staten Island Chamber of Commerce exposition.[17] The WPA's arts division also included zoo work in its publicity.

Zoos were an ideal advertisement for New Deal projects because they were free (or inexpensive), wholesome entertainment for the masses and offered appealing visuals and storylines. New Dealers liked to justify investments in recreational facilities on the grounds that these facilities gave the people something constructive to do with the "enforced leisure" created by the Depression. Zoos, designed to please young and old alike, were a low-cost, family-oriented leisure experience. In the story of each zoo's development, construction, and reception, we can see how the New Deal's extensive reach into local politics was "sold" and defended. In each case, New Deal resources made possible what the local communities, even in times of relative prosperity, had been unable to accomplish on their own. Criticisms of the New Deal

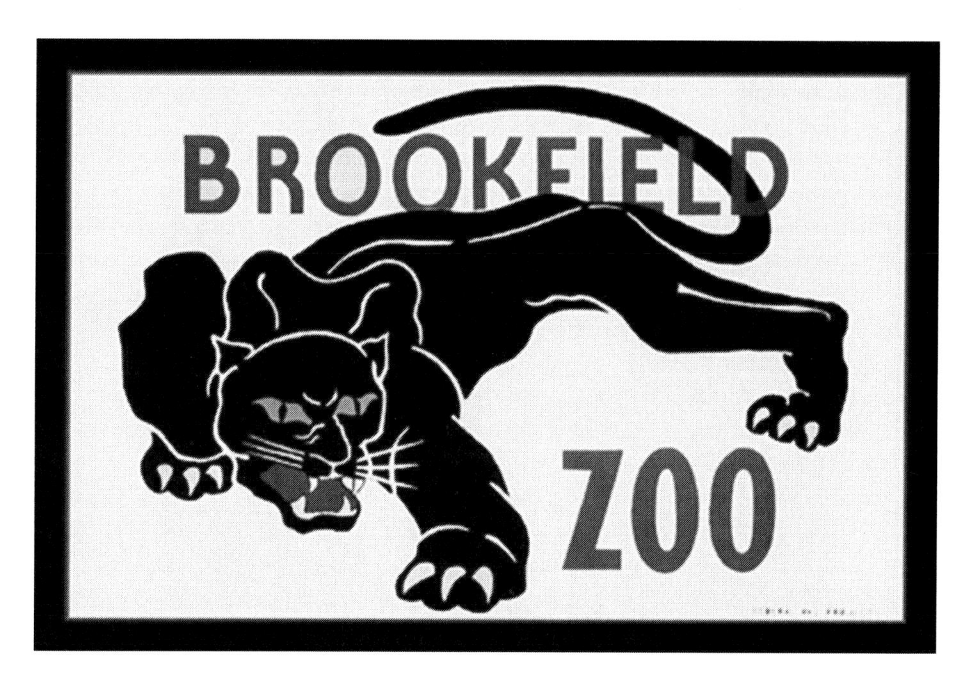

A Federal Arts Project poster advertising the Brookfield Zoo (Chicago, Illinois). FAP artists made similar posters for other zoos. Because nearly all zoos were free, New Deal expenditures on them were easily justified as providing a significant public benefit.

were brushed aside with jokes or deflected with a barrage of publicity about the new zoos' educational and recreational benefits. In the context of economic suffering, free zoos offered a lighthearted stage on which politicians could play to the masses, leading to press coverage that was uniformly laudatory rather than critical.

Overwhelmingly, zoos were represented as designed for and benefiting the masses. Some portion of that benefit accrued from the relief employment provided by zoo construction. The greater value, however, touched all of a city's citizens. Politicians, zoo directors, and local boosters pointed out that a good zoo could draw out-of-town visitors and their cash. An established recreation philosophy justified the expansion of government into an arena that might have been left to the private sector. The low-brow accessibility of zoos combined paradoxically with their presumed scientific orientation to make them simultaneously entertaining and educational. And the American people *were* entertained, in numbers that that dwarf modern zoo attendance figures. But zoos also served a less obvious purpose of keeping the masses under control. Just as relief employment programs were justified in part as measures to defuse radical political and economic movements, so too zoos promised to keep the unemployed busy and happy, literally getting them off the streets and into a space where they could be supervised. And, it turned out, some of them had much to learn about proper public behavior. Thus, the New Deal zoo both served and disciplined America's lower classes.

The St. Louis Zoo and the Will of the People

Zoo advocates pitched the institution as one whose appeal transcended class and other divisions in society. A historical overview of Seattle's zoo, written during the New Deal's final years, asked rhetorically, "What classes of people view these animals at the zoo?" The answer was that "people from all walks of life"—young and old, hunters and conservationists, elementary school children and university students, farmers and merchants—"benefit from the zoological collection, for the zoo is a gathering place for all classes of people."[18] This line of argumentation was developed quite effectively in St. Louis, Missouri, where taxpayers had voted a special property tax assessment to fund the city's zoo, museum, and library. The St. Louis Zoo was one of the nation's oldest and was a source of pride to the city. When St. Louis struggled with lower tax revenues in 1927, however, some city officials sought to divert the zoo tax to fund basic social services for the poor. A vigorous defense of the zoo was launched in response, with advocates arguing that of all three of the city's cultural institutions, it was the most important because it most obviously benefited *all* of the city's citizens. St. Louis was one of only a handful of cities that funded zoos through a special tax, but the arguments offered on the zoo's behalf were echoed across the nation during the next decade as communities debated whether to use New Deal programs to benefit zoos.

After World War I, St. Louis citizens voted to fund the city's zoo, library, and art museum with an eight-cent property tax per hundred dollars of assessed value. Two cents of this tax were earmarked for the zoo. The tax measure passed thanks in part to a "vigorous organized campaign" involving school children to persuade "voters that St. Louis should have a thoroughly equipped and stocked zoological park, for the instruction and entertainment of the people."[19] Parading children with placards had been utilized in other cities' zoo tax campaigns, but not always with St. Louis's success.[20] The zoo served adults as well as children, however, as a contemporary editorial made clear: "Nothing that St. Louis has done has better expressed the high ideals, aspirations, intelligence and culture of St. Louis than the voting of independent incomes for the support of these institutions. Nothing has gained for St. Louis greater credit in the estimation of the people of the United States and elsewhere."[21] It was also a measure that made the St. Louis Zoo among the nation's best funded, giving it an annual budget of several hundred thousand dollars. For this price, citizens enjoyed free admission to a large and growing institution with a top-notch animal collection.

In the midst of economic hard times, however, not all of St. Louis's politicians were convinced that a free zoo was of the highest priority. Indeed, they wished to divert some or all of the special tax for other purposes. After a legislative attempt in state government to repeal the tax died in committee, the mayor and his allies on the board of aldermen simply announced in 1927 that they would be applying some or all revenue from the special tax for other, general municipal purposes.[22] Together the tax raised about $900,000 that the aldermen wanted to use to address a budget deficit.[23] The mayor's announcement set off a fierce political debate about how best to use scarce public resources when poor people were suffering.

On one side of the debate, zoo officials emphasized that the revenue reduction would cripple their operations. Hoping to get the funding back without going to court, George Dieckman (the president of the Zoological Society and the vice president of the Zoological Board of Control) and George Vierheller (director of the zoo and secretary of the Zoological Board of Control) appealed to the aldermen at an August 1927 meeting. Veirheller told them that because they had just recently spent money on a new reptile house, they had only around $135,000 left in the budget, not enough to feed all the animals for the year. He asked for an additional $100,000 more to make it through.[24] Dieckman warned that losing the funds would be "disastrous for the zoo," a prediction that he repeated over the next few months. From his perspective the zoo was a most valuable "civic asset" that should be protected.[25] Both men begged for at least a smaller cut in their expected budget.

Veirheller and Dieckman's request resonated with several aldermen who offered populist support for St. Louis's zoo-going citizens. Although these men were not willing to give up all of the special property tax assessment, they championed the "people's" zoo over the elitist art museum and library. Alderman Schwartz "declared loudly he favored giving the zoo all it needed, but as for the Art Museum and the Library, they had enough already. He said the museum was a fine thing, but he guessed most of the pictures were imitations anyway. There was no question about the monkeys and elephants being real." Alderman Watts complained that the "library's got too many books already," they were "piled clear to the ceilin' ... and nobody reads 'em anyway." Instead, he "delighted in making visits to the Zoo," where presumably the rest of St. Louis's non-bibliophiles felt more comfortable too. These anti-elite sentiments were fully supported by Dieckman and Veirheller, who told the aldermen that around two million people visited the zoo annually, many "of them strangers to the city who were greatly impressed." As the meeting progressed, the mood seemed to be changing in the zoo's favor. A few aldermen "didn't care so much what happened to the Museum and the Library, but they would see that the Zoo got its money."[26]

The budget hawks, however, were not ready to concede defeat. The mayor and his allies stuck to the view that taxpayer money would be better spent on services for the truly needy *people* of St. Louis rather than on its exotic animals. During the aldermen's initial discussion over funding, the assistant comptroller quietly slipped out of the room to retrieve his boss. Comptroller Nolte returned to reiterate that not only would the city shift the eight-cent tax into the general fund, but also that he would not "guarantee" that the zoo would receive the appropriation it requested. Nolte cast the zoo as an unnecessary luxury, arguing that it had spent money on "baths and tile houses for monkeys when we haven't place in our institutions for the sick and invalid."[27] Human needs had to take priority over animal needs. This was a familiar argument, but one that the zoo's defenders were prepared to answer.

The St. Louis newspapers came to the zoo's defense, casting the conflict as one between the "people" and the "politicians." The *St. Louis Star* told readers that "the city administration" found itself "confronted with a deficit" and resorted to "this high-handed betrayal of the people in order to obtain control of the revenues and reserve of these institutions and divert them to general uses." "Obviously," the paper editorialized, it was a mistake to remove the revenue because doing so would lower the standards of the institutions and "deprive the people." The St. Louis Zoo was a non-partisan institution because of the special tax. Repealing the tax would put the zoo's future back in the thick of party politics, called "City Hall and its interests." The paper questioned the legality of the city's action: "these mandates of the people have been swept aside in a most summary proceeding, without notice and with a hearing. Can this be done legally? Is it possible that a formal edict of the people of this City of St. Louis can be repealed by the Board of Aldermen without the authority of the people?" City Hall took the position that the special tax was itself unconstitutional, but the *St. Louis Star* saw only "an arrogant and unprecedented defiance of the expressed will of the people of St. Louis [that] should be vigorously condemned by the forces of public opinion." The effort to take the zoo's money was "vicious in its implied attempt to ignore the will of St. Louis taxpayers and voters."[28] The zoo was clearly a "people's" institution, and it had to be protected against the politicians who pretended to speak for the people, but actually acted on their own whims.

Given the mayor's position, however, Dieckman and Vierheller knew that their only hope rested with the courts. The mayor claimed that the tax was unconstitutional, but he had no interest in testing his legal theory before an actual judge. Dieckman and Vierheller consulted with a lawyer who advised them to ask for the money and when refused by the Board of Estimate and Apportionment, sue "in the Supreme Court for a writ of mandamus compelling payment."[29] The Zoological Board of Control, however, was not a citizen board, but rather was made up of the exactly the people who were trying to take the money: the mayor, the comptroller, the Board of Aldermen vice president, the park commissioner, and the president of the Board of Public Service.[30] Because of its political composition, the board would not initially vote to challenge the tax diversion on the zoo's behalf. After the library brought suit independently, however, the zoo board relented, with the mayor and comptroller gambling that a favorable ruling (i.e., against the tax) would give them greater control of the budget.[31] This case went to the State Supreme Court in Jefferson City, where the defendants (the mayor, the comptroller, and the president of the Board of Aldermen) argued that the taxes were unconstitutional because they were special and local. Presiding Judge J. T. White, however, ruled that they were constitutional because voters had authorized the taxes. The judge concluded that simply because the city believed that other parts of the municipality needed the money did not mean that they had the authority to transfer it there. He decided that by doing this they were wrongly claiming that they could "substitute their judgment" for the "express approval given by the taxpayers."[32]

When the Missouri Supreme Court ordered the St. Louis Board of Estimate to pay the additional two-cent tax to the zoo, the *St. Louis Star* reminded readers that it had "steadfastly" opposed that diversion "when it was first proposed" and in the months thereafter. The paper cast the court decision as a victory for democracy, noting that the city's action had been "thoroughly unpopular."[33] And for good reason, since the "people have pride in their Zoo, Museum, and Library. They afford instruction and amusement to thousands of St. Louisans and St. Louis visitors every day."[34] The *Star* had always acknowledged that the budget grab had illuminated the financial plight of the city, but maintained that it was wrong to take money set aside for special purposes. From the beginning of the controversy, the paper had suggested "seeking revenue sources elsewhere, as has been done in other cities." They noted vaguely that cities like Detroit and Cleveland were "carrying on" their work "without corresponding revenues," and while admitting that they had no clear recommendation about how to raise revenue, insisted that it was "a problem for city officials to solve."[35]

But city officials continued to see the zoo as a piggy bank to rob, especially after the Depression officially hit. In 1932, facing a $652,000 deficit because of expenditures for relief purposes, the mayor took a different approach by seeking a citizen referendum on one of two proposals to reduce the zoo's revenue. The problem, the mayor explained, was that the city charter forbade lowering taxes. It stipulated that existing taxes must be abolished before lowering or raising them. His plan, he claimed, was to use a citizens' referendum to siphon off some of the eight-cent assessment because the revenue of the zoo and art museum were "greatly in excess of their needs in these times of depression, especially in view of the fact that the proceeds" of the zoological tax had increased from "$115,908 in 1917 to $265,664 in 1932." "Accordingly, it was decided by a majority vote" of the Board of Estimate and Apportionment to "reduce the revenue of the said institutions to a lower level, in harmony with the general policy of economy that has been applied to all other departments in city government."[36]

Once again, the mayor's action met with fierce public disapproval, this time colored by the new hardships of the Depression. While the mayor argued that it was unconscionable to spend money on a zoo when the people were suffering, his opponents argued that it was necessary to fully support the zoo precisely *because* people were out of work.

The first line of defense held that the "enforced leisure" created by the Depression made institutions like zoos a crucial "safety valve" to occupy people's time. The pressures of unemployment made the zoo, which was free, all the more important to ordinary Americans. One pro-zoo editorial pointed out that "at a time when so many people have leisure both the Zoo and the Art Museum enjoy a greatly increased attendance." Americans who "are out work for the first time in their lives have time to spend there," and the two million annual zoo visitors are not "chair-warmers" but have "a *bona fide* interest" in its collections. The increasing attendance revealed that "there may never be a time when these things were more needed" or a worse time to cut their special taxes.[37] Another editorial asked how, if the zoo was "crippled" by an inadequate budget, "is the enforced leisure of thousands to be made other than a wasted leisure?" This writer wondered darkly whether "time, weighing heavily on many hands" might "be devoted to much less profitable, not to say antisocial, uses than to recreation and instruction?" The zoo served a particularly "special and important purpose" in this regard because it had "an appeal to a class even more numerous" than the "ambitious and intelligent class" that visited the art museum.[38] In other words, members of the lower classes would spend their time in the zoo rather than on the street or at home, where they might cause trouble of some kind. Another writer was more explicit. Ely Smith, the president of the Council on Civic Needs, viewed the zoo as a "means of crime prevention" particularly "in periods of depression."[39] Who wanted those two million annual visitors turning to crime?

A second Depression-related angle emphasized the economic benefit to the city of a zoo that attracted visitors from other states. Zoos had always been touted as symbols of civic pride, but now they were pitched as magnets for out-of-town money. One civic leader feared that reducing the zoo's appeal would hurt the city "in its trade territory."[40] An editorial warned that "an assault" on this local attraction, "in which tens of thousands of tourists revel as well as our on population, will make St. Louis poor, indeed, will render it a less desirable place either to visit or to live in, will neutralize effort in other directions for bringing to us worth-while guests."[41] Another writer argued that the modest zoo tax "brings thousands of visitors to St. Louis every year," a pretty good return on the investment.[42] Much as modern sports stadium supporters claim large multiplier effects from public investment, so too zoo advocates stressed the economic activity generated because of the zoo: "The money that now goes to our Zoo and Museum results in a greater return to St. Louis than in any manner it possibly could be used. It not only provides pleasant and educational entertainment cheaply to each and every St. Louisan, but it brings within our gates thousands of visitors who send money which goes to St. Louis hotels and merchants. It advertises St. Louis to the world in a manner that cannot be duplicated. It results in attracting wealth and industry, for in all ages the cities most advanced in music, learning, and arts have always become the most prosperous and renowned."[43]

Despite the economic conditions, tourist travel during the Depression was normal or even better than usual, and after 1933 expanded greatly because of the New Deal's construction of roads, bridges, rest stops, and parks. Americans had somewhere inexpensive to vacation and, as car ownership rose, a way to get there. The result was that a city zoo could draw more than just local residents. For the summer of 1929, for example, Dr. Ray Hal of the finance and investment division of the U. S. Department of Commerce estimated that tourists spent 5 million dollars at roadside attractions and camps. He predicted that tourist travel should be the same in 1930 "despite the stock market deflation of last Fall." In fact, better roads and more cars would actually stimulate travel, with positive results for the economy: "Whether these tourists journey within or across the border lines, the purchases made en route for supplies, food, accommodations, etc. will add to the national income...."[44] For cash-strapped city governments, a well-appointed zoo just might attract some of these visitors and their money.

To support the tourism thesis, George Vierheller penned a substantial article for the *St. Louis Star* about his zoo's fame and its contributions to the city's economy. He claimed that it was "the world's finest Zoological garden." Its fame spread as news reels featured the zoo, "stories in magazines have been followed by other stories, and best of all," satisfied visitors made personal recommendations to their friends—"Going to St. Louis? 'Well, don't miss the Bear Pits in Forest Park: they are wonderful." Vierhellar claimed that nearly 40 percent of the zoo's visitors were from out of state, a statistic derived from a count of license plates on parked cars; one late-summer count revealed vehicles from 23 states. Adding animals and building new exhibits ensured that the zoo stayed in the public eye, attracting new visitors and justifying the public expenditures. "Stories and visitors go out, and the visitors come in. Travelers add another day to their trip and St. Louis profits thereby."[45]

Similar arguments were made in Seattle, San Antonio, New Orleans, and other cities. The Seattle zoo was described as a destination for out-of-town visitors who then spent money "on the street cars, boat rides, theatres, restaurants, gas stations, department stores," and so forth. It was a "proven" fact that "for a drawing card, the zoo cannot be equaled."[46] As supporting evidence, Seattle's zoo director, Gus Knudson, wrote monthly reports to the Park Board that not only listed the various states from which visitors had been drawn, but also named individual out-of-state patrons and described their travel habits and motives for visiting Seattle.[47] In New Orleans, the zoo superintendent also kept track of out-of-state license plates in the parking lot.[48] San Antonio's mayor believed that "an increase of public interest in the municipal zoo would materially assist the city in its effort to keep this feature of the parks 'ahead,'" and he declared himself to be all for "'a bigger and better zoo.'" As he reasoned, "'anything that will help keep San Antonio first has my endorsement. There is no doubt but that a larger zoo would increase interest among tourists, and I am in favor of doing everything we can to this end.'" A local editorial echoed this view, describing San Antonio's zoo as a "tourist and public asset."[49]

Back in St. Louis, opposition to the mayor's cost-savings proposal rose and the issue became one of civic pride. "Cutting these small taxes, or forcing the Zoo and Art Museum to depend on what politics hands them for expenses means their final decline and disappearance as nation-wide advertisements for St. Louis." Sinister politic motives with catastrophic civic results were detected: "They are taking advantage of a period of temporary hard times to get political control over the Zoo and Art Museum which they tried and failed to get a few years ago. What we would have under such arrangement is exactly what Detroit now has, with its public library open only half time, its art museum shut most of the time, and its Zoo open only in summer and on an admission fee basis, simply because the City Hall runs the finances of all three."[50] Erwin O. Schneider, the editor of the *St. Louis Star and Times*, wrote: "The newspapers have again called attention to organized agitation by certain members of the city administration to financially hamstring our Zoo and Art Museum." The tax costs of the zoo and art museum were "negligible" for property owners and zero for renters, and all citizens recognized the "civic need" for these institutions. "Just imagine taking your friends and relatives from out of town to Forest Park and having to tell them that our Zoo and Museum are no longer what they formerly were because we permitted our politicians to divert the funds necessary for their present high standard! It was not so long ago that our Zoo consisted of several iron cages filled with a few sad-eyed, moth eaten bears. Can any progressive St. Louisan desire to go backwards?" No. The newspapers and civic organizations had a duty to "arouse the citizenry" against any efforts to "disable" or "destroy" these important institutions.[51]

The ultimate fear was that St. Louis would lose its zoo altogether. Supporters pointed out that it might be impossible to resurrect a crippled or dead institution. Without the entire tax assessment, the zoo would not be able to maintain its "full vigor," and if the tax were lost entirely, "it would destroy" the zoo. "It must be born in mind that if we do cripple or destroy the incomes of these institutions because it is hard to support them through a business depression, it will be difficult to restore support to them when the depression ends." Both taxes were voted only after years of effort upon the part of friends of the institutions. "The proposal contemplates a backward step for St. Louis. We will have to do better than that if we are to make any pretension of having a city of the first grade. A city of the first grade would not think of crippling or destroying either its Zoo or its Art Museum. What Mayor Miller and Mr. Nolte propose is that we content ourselves with having a big Indian village."[52] "Private benefactors die or go broke. Both these contingencies have crippled the Cincinnati zoo. In St. Louis the sponsors are the entire public which gladly approves the apportionment of a small part of its tax money to give St. Louis one of the several institutions which have made this city famous."[53]

The newspapers claimed to speak for the voters and the public interest. The *Post Democrat* refused to "believe the people will consent to [the zoo and art museum's] destruction; we doubt if the people would consent to diminish their revenues."[54]An editorial in the *Globe Democrat* argued that "It is unthinkable that the voters who approved those special taxes many years ago and who have supported firm resistance against hard-fought attempts to divert funds so assembled will now turn around and approve reductions, even in the time of present need." The mayor wanted to put the issue before the voters, but the *Globe Democrat* urged him not do so, and it implored voters, if an initiative proposal were offered, to defeat it.[55] Unlike the "people," the city controller's "ideas of city government do not go outside the bare routine — water supply, paving, police and the like. The cultural and recreational services of municipal government — which must be paid for with tax money if we are to have them at all, do not appeal to him."[56] Even — perhaps especially — during an economic depression, the people valued public cultural and recreational institutions.

The story of St. Louis's political battles over public zoo funding captures the range of arguments made to justify taxpayer support of this institution. The debate ended on the eve of FDR's national election, and the defeat of the mayor's proposals happily coincided with the establishment of federal programs that would allow St. Louis to increase its already generous support of the zoo.

The New Deal entered the national scene in the context of debates much like that in St. Louis, and it provided a solution that satisfied both sides. For those who wanted to provide relief to the poor and unemployed without expending large sums of local tax money, the New Deal's relief labor programs were an acceptable solution. At the same time, those relief laborers would have to be occupied doing something useful, and what could be more beneficial to a city and its citizens than a new or renovated zoo that would polish the city's civic reputation? The pro-zoo arguments made during St. Louis's zoo funding crises thus moved to the national stage with the arrival of the New Deal.

Work Relief for Ordinary Americans

The immediate impetus for FDR's alphabet soup of federal programs was unemployment. Millions of Americans needed jobs or they would be homeless and hungry. Each new zoo may have housed and fed animals, not people, but its construction typically employed hundreds of skilled and unskilled laborers. Zoos may not have been necessary, but jobs were. Still, relief labor programs were essentially welfare programs, and the workers were branded with the same kinds of stereotypes and criticisms that coalesced around later generations of welfare recipients. Relief labor programs, although popular with the unemployed, attracted criticism from some quarters on the grounds that the laborers did too little work and too much loafing. The Tulsa Zoo director expressed this sentiment by joking that WPA stood for "We Piddle Around."[57] When a Central Park Zoo bear, Mishka, escaped from her cage while the zoo was under reconstruction, reporters gleefully claimed that the relief workers "moved faster than they have in weeks."[58] These comments were offered good-naturedly, however, with a kind of wink-and-nudge that acknowledged some work relief inefficiencies but appreciated the end results. On balance, the efforts of these workers received praise from local officials.

Representing the less charitable view of relief labor was New York City's park commissioner, Robert Moses, whose dismissive attitude toward the armies of unemployed working on his hundreds of public works projects may well have reflected his class bias. Moses was a paradoxical character. He sought to build an enormous infrastructure — parks, roads, bridges, etc. — for the benefit of the "common man"; at the same time, he frequently expressed scorn for the abilities and work ethic of the men employed by the relief agencies. In part, his criticisms stemmed from a deeper philosophical opposition to how these agencies operated. Moses disdained the local work-relief activities that cities had implemented in the late 1920s to give unemployed men jobs in the parks doing "petty chores hardly distinguishable from maintenance."[59] In this sense, he actually shared the view of FDR's work-relief administrators who distinguished the construction-focused nature of their programs from the jobs "shoveling snow or raking leaves in one of the public parks" that cities provided "in lieu of *planned useful* work."[60] The public benefit of the old-style work was minimal, and even its relief value was limited. In 1930, the Manhattan Department of Parks had nearly 2,000 relief workers assigned by the city, though they worked only three-day shifts.[61] But Moses was not particularly enthusiastic about the New Deal system either.

Although Moses proved quite adept at wringing resources from the federal and state agencies, he complained bitterly about the inefficiencies of the relief system in terms of its ability to support public works projects. True, he echoed federal administrators in believing that "work relief is better than home relief and civil works is better than work relief." However, he wished that federal funds would allow "contract work" (i.e. awarding construction projects to private contractors), he chaffed at "needlessly stupid rules and regulations," and he felt that a larger percentage of funding should be "allocated for materials and equipment."[62] When asked by an aldermanic committee investigating relief in 1935 whether he was "in charge" of work relief in the parks, Moses replied, "to the extent that is possible to be in charge of anything under the Federal, State and local relief rules."[63] To demonstrate the inefficiency of these rules, he described how they had hampered work on the Central Park Zoo. The FERA regulations dictated that workers could be employed only until they had earned enough to meet their minimum budgetary needs for a month, an amount dictated by the state relief agency; yet, the hourly wages of skilled laborers in New York City were determined by prevailing union wages. So, some workers might make $12 a day but be capped at $60 a month, effectively limiting them to five-day shifts. Moses described how this "absurd" situation affected the carving of bas reliefs at the Central Park Zoo: "we had a case where stone cutters were to work on bas relief work in the Zoo, and it would have figured out that one man would start working on the lion's tail, and when he got half way through, another man would do one foot and another man would do another foot, and about the time the third man got through working on it, the lion would be finished."[64]

Moreover, while official reports from the FERA and WPA promised that relief laborers put in a hard day's work, Moses did not fully support that proposition. He disparaged the quality and dedication of relief workers. He noted that among the unskilled laborers assigned to him, a fair number had no work experience at all, another subset were physically limited in their abilities, and still another group "will not only not work themselves, but . . . feel it is their bounden duty to see that nobody works."[65] He complained to the mayor that the WPA had assigned "numerous vagrants" as relief workers to the Parks Department.[66] Moses estimated that from this motley mix of workers he was fortunate to get "50 or 60 per cent efficiency, which is the most you can possibly get under these Federal and State rules."[67] In a light moment with reporters, he publicly joined New Deal critics, joking that he could easily fill the Prospect Park Zoo's 230 empty cages "four times over" with "boondogglers."[68]

A year later at the Staten Island Zoo's opening ceremonies, Moses reiterated his fundamental distance from the working class. In his version of the zoo's construction, he cast himself as a heroic park commissioner battling the ineptitude and inefficiencies of the work relief system. Not even mentioning the hard work of hundreds of relief laborers, he instead reviewed the "difficulties and trials and vicissitudes" that he had personally overcome to shepherd the zoo project "through all the different alphabets"— the FERA, CWA, and PWA. Such federal programs, he claimed, were "geared to do simple work," not to create fine structures such as the Staten Island Zoo. While the federal government had changed the names and administrative structures of its relief programs several times during the first few years of the FDR presidency, Moses had made the most of his relief workers, completing 1,700 parks renovation projects in just six months after taking over as park commissioner on January 19, 1934. This achievement was made possible by driving park and relief employees to adopt his work ethic. Moses's comment about "simple work" reflected his frequent complaint about relief workers paid to do little more than stand around or rake leaves. To avoid such slacking, Moses sought "the toughest" foreman and field supervisors that he could find.[69] Even so, he noted the difficulty of locating "skilled workmen to do the work required in such a project."[70]

Most zoo directors, however, defended the relief workers. When Woodland Park Zoo Director Gus Knudson described the accomplishments of CWA laborers in his zoo, he emphasized the tough physical strain of their tasks and he challenged any park commissioners who heard rumors of WPA laborers "laying down on the job" to "come and observe for yourselves just how the work is getting on."[71] In Chicago, Director Ed Bean offered similar praise for the CWA and WPA crews that spent several years getting the Brookfield Zoo into shape. In report after report between 1934 and 1939, he detailed the progress made by laborers, painters, skilled craftsmen, and artists.[72]

Moreover, despite Moses's sniping, these supposedly inefficient laborers built zoos with amazing speed. The Central Park Zoo opened to the public just eight months after construction began, in no small part because the New Deal relief programs emphasized mass employment, meaning that many hands, however unskilled, were available for labor. The actual construction utilized a veritable army of laborers working literally around the clock. Nearly 700 men worked a day shift, and then another 400 took over for a night shift made possible by high-powered carbide lights.[73] The local press and civic boosters were pleased, offering nothing but praise for these accomplishments. For example, at a well-attended Staten Island "Exposition of Progress" sponsored by the Chamber of Commerce, Richmond Borough President Palma touted the "absence of 'boondoggling' on Island WPA projects," and as if to demonstrate what had been achieved, the Staten Island Zoological Society entertained Exposition goers with live snakes and animals from the zoo's growing collection.[74]

The Central Park Zoo's opening day was marked by speeches from the city's political leaders, all of whom took the opportunity to defend New Deal relief programs. Three local radio stations broadcast the opening ceremony.[75] Although a cold absented Moses, Mayor La Guardia, former Governor Alfred Smith, Welfare Commissioner William Hodson, and the Temporary Emergency Relief Act State Chairman Alfred Schoellkopf all spoke. Smith, escorted to the stage by 300 children, reminded the 15,000 spectators that the zoo had been completed in record time, and he praised it as "one of the finest of its kind in the world." Schoellkopf described the zoo as "an illustration of the kind of work that can be accomplished under an intelligent and well-planned relief program." Hodson suggested that Moses next build "an institution" where the city could put the "crackpots" who did not want to pay taxes for relief work. La Guardia echoed Hodson by claiming that "the more they [the "crackpots"] abuse us, the more we'll do."[76]

A year later, Al Smith echoed this general defense of the New Deal at the opening of the Prospect Park Zoo. Setting aside his animosity towards FDR, he played cheerleader for the New Deal, contrasting what "relief funds" had "accomplished" in the Central Park Zoo with the wasteful use of privately managed relief workers under the previous mayoral administration.[77] Then, in a nod to the local political significance of the New Deal, he praised the mayor for "securing the funds and in giving at the same time employment to some part of the large army of unemployed who have been suffering, through no reason of their own, over a period of five years."[78]

Soon after Moses formally announced in March of 1934 that the Staten Island Zoo project would move forward, one letter to the editor from a nature enthusiast stressed that the zoo would be "a valuable asset to the educational facilities of Staten Island" and welcomed the unemployment relief to be provided by its construction. Another letter writer praised Moses and argued that "no CWA funds could be better expended than for this building which will provide just those recreational facilities — for adults as well as for boys and girls — which President Roosevelt and other leaders of today have strongly indicated the future will require."[79] Here the writer referred to a favorite idea of New Deal architects — that properly directed recreational activities would elevate the physical, intellectual, and moral health of the masses. Across the nation, men from the laboring class built zoos that would be used by working-class families. These zoos would be largely free, ensuring that they were open to all. Civic leaders extolled the civic virtues of giving the un- and under employed a socially acceptable way to take advantage of their enforced leisure. Politicians and zoo directors spoke eloquently about the educational value of zoos.

Leisure and Recreation for the Masses

The development of zoos paralleled the growth of America's local park system during the 1920s. In the final five years of that decade, the National Recreation Association calculated that 38 percent was added to the nation's local park acreage and the number of counties with a park more than doubled.[80] During the New Deal, many more parks were added, and most of the existing ones improved. The WPA's final report listed work on 8,000 parks ("new or improved"), in addition to tens of thousands of playgrounds, ice rinks, baseball fields, and other facilities constructed in parks.[81] This intensive attention to public parks satisfied the demands of recreation advocates and park professionals who had spent the previous decade articulating the larger benefits society derived from expansive recreational activities. For these advocates, access to recreation was both a fundamental right and a social necessity. It gave all citizens an equal chance to enjoy their leisure time, and it cultivated habits of mind and body that led to better citizens. Zoos fit easily into this paradigm, with their advocates stressing both the wholesome, free entertainment they offered and their educational role.

The official reports for New Deal work relief programs treated recreational infrastructure improvements (playgrounds, parks, swimming pools, sports fields, zoos, and so forth) as a distinct category in their tables, graphs, and narrative chapters. Providing recreational opportunities for Americans was presented as being of national importance. The FERA emphasized that before its parks building program, "the children and young people of our American cities had little or no place to play except in the streets."[82] The WPA, justifying the 11 percent of total project funds put towards parks and recreation, argued that "local officials throughout the country recognize the growing problem of leisure time in America."[83] Even before President Franklin Delano Roosevelt took office, the National Recreation Association had touted the "unusual contribution which parks are able to make in the present period of unemployment" when so many could not "afford the commercial types of recreation," and it stressed the ability of parks to provide employment for relief labor.[84] Harry Hopkins claimed that demand for recreational facilities was so strong that when the CWA shut down, "workers, anxious to have their children benefit from these places to play, asked permission to finish their jobs without pay." The New Deal, Hopkins concluded, enabled "leisure, once the privilege only of the rich" to "now belong to everyone."[85]

Historian Phoebe Cutler agrees with this assessment, arguing that the New Deal had a profound impact on public recreation, funneling enormous resources into the development of a new kind of park, one that "eliminated the leisure gap" between the genteel and the general public. Frederick Law Olmstead and other nineteenth-century landscape architects had conceived of parks as naturalistic environments in which one could pursue quiet, relaxing activities; in contrast, the National Recreation Association (established in 1906) promoted the conversion of public parks into spaces where citizens could engage in sports and play-oriented activities. Olmstead objected to zoos because they brought huge, noisy crowds into parks and disrupted their naturalistic design; in contrast, the public recreation advocates embraced zoos precisely because they gave the public, especially children, a way to use public parks.[86] And, fundamentally, the recreation movement's rhetoric of equal opportunity meshed neatly with President FDR's liberal ideology. The National Recreation Association's slogan nicely captured this rhetoric: "That every child in America shall have a chance to play. That everybody in America, young or old, shall have an opportunity to find the best and most satisfying use of leisure time."[87]

Robert Moses was a leader in this recreation movement, and it is easy to see how his three zoos were designed to provide recreational opportunities to New York City's residents. As he enthused in an *American Magazine* article in 1938, "I believe in bigger and better construction for public recreation because I am satisfied that it makes better people. This is not an endorsement of government extravagance, but simply a plea to do adequately and well what a community needs to make it really livable."[88] Around the same time, he defended the high cost of park improvements by arguing that "recreational facilities must be accepted as legitimate burdens" which were "necessary for the health and well-being of the city." Moreover, their benefits came not only in immediate recreational opportunities, but also from their positive impact on real estate values and their "dollars and cents" impact; in other words, park building stimulated economic health.[89]

In the 1920s and 1930s a city zoo provided recreation as well as a space for a broader range of public activities. For example, when Seattle's Woodland Park Zoo purchased its first elephant, "Wide Awake," from the Singer Midgets circus with $3,500 in spare change donated by 25,000 children and adult residents, the city organized a day-long celebration to welcome the elephant to her new home. The day began with a bicycle race. Then Wide Awake, wearing an elegant headpiece and a purple velvet caparison decorated with rose and gold embroidery and tassels, paraded several blocks with the chief of police, the mayor, members of the Park Board, employees of the Park Department, and several hundred children, many riding decorated bicycles or leading Shetland ponies. After a formal christening ceremony, the festivities continued with pie and chocolate-eating contests and athletic events.[90] Such pageantry brought a community together and suggested that a zoo's public role extended far beyond its gates. A zoo could focus and make evident the bonds between a city's citizens.

In November of 1929, just a week after the stock market fell, a record-breaking crowd of 75,000 gathered in San Antonio's Brackenridge Park for the opening ceremonies of the zoo's new barless bear and monkey exhibits.[91] Local papers claimed that this was the largest daylight crowd the city had ever witnessed. More important, the crowd represented the whole of the citizenry, "not just the children" and their parents, but also the "very old." Even those who had been skeptical about the necessity of zoo improvements were swept up in the "pleasure, happiness, and national advertising" of the opening events[92]

As the Depression worsened, zoo attendance soared. George Vieheller announced to the American Association of Zoological Parks and Executives that the "depression is not apparent in zoological activities" Instead, zoos hit "new records" for attendance.[93] Clearly, New Deal zoos could attract crowds, as evidenced by the success of the Central Park Zoo. On opening day, 20,000 New Yorkers passed through its gates; in its first spring, the Central Park Zoo reported 207,000 visitors in a single week, with 84,000 counted on Sunday alone. By the end of 1939, the zoo claimed that 20 million people had visited the zoo.[94] Opened in July 1935, the Prospect Park Zoo averaged 100,000 visitors a week through its first summer.[95] The Staten Island Zoo, drawing from a much smaller population base, nonetheless attracted 280,107 visitors in 1938, hosted 179 nature "hobby" group meetings, and conducted scores of lectures and other educational activities.[96] Nearly 5,000 New Orleans residents visited their new zoo on a Labor Day weekend in 1939.[97] In Seattle, where the zoo did not keep formal attendance numbers, an estimate of 900,000 a year was given.[98] In 1935 the FERA reported on its range of zoo work, singling out the Brookfield Zoo for special notice because it hosted two million visitors a year, including 400,000 children.[99] The children were particularly significant because they represented the Depression's innocent victims. No one could accuse them of being lazy. The public works programs were giving them a well-deserved place to enjoy themselves and escape the stresses that may have been affecting their parents. The fact that 400,000 children had visited the Brookfield Zoo in a single year implied that the need for such recreational facilities was great.

When local government would not adequately support the zoo, directors turned to the popularity argument. In Seattle, Zoo Director Gus Knudson chided his Park Board for a budget that barely covered maintenance and for not proposing a bond issue to benefit the zoo: "I believe that the places where the masses of the people go are the places to be kept up. The zoo has always had the smallest appropriation while drawing the mass of the people."[100]

Zoo advocates waxed eloquent about the social and educational effects of zoos on ordinary citizens. When the Prospect Park Zoo opened, Al Smith was on hand to extol the public benefits of education and recreation offered by zoos. In the grand scheme of public works, he argued that "I would just as lief close up the Erie Canal as I would deprive the children of this city of the education that comes from a zoological garden such as we have here today."[101] This kind of rhetoric was fairly well-established — international animal dealer Ellis Joseph in 1920 described the city zoo as "one of the principal mediums of innocent amusement" available to the masses — but it took on new relevance and urgency during the Depression.[102] The Central Park Zoo was touted as "a great service" to those "who could not afford the time nor the carfare for a visit to the Bronx Zoological Gardens."[103] The Audubon Park Commission reminded the public that for those New Orleans residents who could not afford a ticket to the circus, "every day is circus day at Audubon Park."[104]

The infant Dallas Zoological Society argued in 1922 that the city needed a "first-class Zoological Garden" because "there is nothing quite so interesting, educational, and entertaining to the children of any City" and "the best is not too good for the children of Dallas."[105] By 1930, the justification had broadened somewhat, and the director of parks urged the enlargement of the zoo on the grounds that "a fine Zoological Garden is one of the greatest educational institutions outside of the school room," is "one of the most important recreational features that any city can furnish for its citizens," and is "one of the best advertising mediums a city can have" drawing outsiders to Dallas to spend their money.[106] Here the emphasis is on providing the "best" for the city's citizens. The proposed educational benefits derived solely from untutored viewing of the animals; only a few zoos of this era had formal education departments, and the Dallas Zoo was not among this group. Rather, civic leaders put their trust in the ordinary citizen's ability to combine a recreational and educational experience at the zoo.

Even small New Deal zoos borrowed the rhetoric developed by their large-city brethren. The Mesker Park Zoo in Evansville, Indiana, (pop. 103,000) attracted thousands of visitors, and despite its relatively modest collection, claimed great effects on the masses. "Of all the recreational facilities possessed by a city there is not one that attracts greater attention or gives more pleasure to young and old than a collection of wild animals and birds, or a zoo as it is more commonly known." The zoo was a "Mecca," offering not just "pleasure" but also "education." Since the Mesker Park Zoo had only a few employees and no formal education program, knowledge about animals was gained through direct observation: " The public learns more about animals and birds in an afternoon at the zoo than by reading half a dozen scientific treatise." And what, exactly, did the public learn? "The sight of these wild animals and birds gives us a sense of the diversity of life, shows us some of the wonders of nature, teaches us to realize the many forms of living things on the globe. Many of these animals are nearly extinct and perhaps our generation may be the last one to see them."[107]

Other zoos made more concrete links between their collections and formal education. In Seattle, the Woodland Zoological Park's director, Gus Knudson, devoted a significant portion of each month's three to five-page report to the Park Board to describing the various teachers and their classes that had come to the zoo. A member of the Seattle School Board took a "keen interest" in the zoo "from the school board's standpoint."[108] In a more general sense, Knudson described how he and the keepers were kept busy answering questions: "how a bear drinks its water, whether it laps or sips; how a deer arises, whether it comes up on its fore or hind feet first," and so on.[109] Even the zoo farm was justified on the grounds that it would be "of great value to [city children], educationally."[110]

Lofty claims of zoos' educational benefits coexisted easily with a populist rhetoric that downplayed the significance of scientific knowledge about animals. At the opening ceremonies for the Staten Island Zoo, former Governor Smith — by now the city's acknowledged headline act at zoo openings — played the role of public comedian, good-naturedly extolling the public benefits of the zoo while humorously calling attention to his own lowbrow interest in animals. The Staten Island Zoo was designed not just to amuse the public but also to provide formal zoological education. The basement included an auditorium, a laboratory, and four lecture rooms in which New York students could study animal biology under an educational program coordinated between the Zoological Society and the city's Board of Education.[111] Smith lauded this education mission, defending the public expenditure on the grounds that "no money was ever wasted on public education." At the same time, Smith diminished the importance of formal education, claiming (tongue-in-cheek) that after a few years in his new honorary position as "curator" of the zoo, he would be "the greatest animal expert ever found in this particular part of the United States," with a knowledge gained not from books, but from "actual experience"—"I know exactly what time every animal should go to bed." Nor was Smith deaf to the entertainment value of zoos. He noted appreciation for the new zoo's reptile collection, but, admitting no real fondness for snakes, hoped the zoo could add some larger animals; jokingly, he promised to "pick up a coupla [sic] lions" during some quiet night at the Central Park Zoo.[112] Smith pleased the crowd and the press, dominating news reports, capturing the sense of gaiety and spectacle appropriate in a zoo space, and looking forward to an even "better" zoo (i.e., one with lions) in the future. Smith was no elitist, and his opening day speeches made clear that New York's three new zoos were intended for the pleasure of the city's ordinary residents.

Smith fully appreciated the child-oriented, populist appeal of zoos. He willingly participated in making New York's zoos fun for children. Soon after the Central Park Zoo opened, it hosted a frog jumping contest, with Smith serving as the master of ceremonies.[113] Smith also sat as the presiding judge in a review of names for the zoo's new animals. Sifting through some 48,000 suggestions offered by the city's children, Smith good naturedly announced the winning names, occasionally substituting one of his own when the children's suggestions were inadequate. No child, it seemed, had thought of "Tammany" for the tiger.[114] The *New York Times* ran six long articles about this contest, suggesting how important this kind of diversion was for a city in the midst of the Depression.

In order for all Americans to be able to take advantage of these diversions, these opportunities had to be low-cost, and in this respect America's zoos satisfied. Only a handful of the nation's 88 public zoos in 1932 charged admission, and of the seven that did, two offered free days each week. Even on paying days, the cost was fairly reasonable, five to 30 cents in an era when a candy bar or a soda could be had for a nickel.[115] Still, two bits was a barrier to many, as could be seen from attendance figures. At Chicago's Brookfield Zoo, only seven percent of its first two million visitors paid admission; the rest all came on free days. Dolores Gordon, a resident of Chicago, recalls "overwhelming" crowds on those days; when she went on a paying day, she felt "wealthy."[116]

The New Deal zoos were not only free, but also gave ordinary citizens access to upscale amenities. Federal relief administrators believed that well-designed work-relief projects could employ Americans with a wide range of skill sets and satisfy a range of public needs. The arts were among those needs able, as the FERA put it, to enhance "our enjoyment of life." The FERA perceived its public art program as "of special significance" and proudly noted that many of the public buildings constructed under the CWA and the FERA "were enriched with a multitude of beautiful art works."[117] The successor WPA similarly touted the role of arts, as emphasized in a 1936 WPA report that highlighted WPA work at the Toledo Zoo to construct a band shell and outdoor amphitheatre as well a natural history exhibit that employed a "staff of artists."[118] The quality of some of this work can be gauged by Robert Moses's retrospective survey of New York City's public statues, in which he singled out one WPA work for particular praise: the "Zoo frescoes in Central Park, [are] simple, attractive, and delightful."[119]

Some zoos received sit-down restaurants of graceful design and tasteful decoration. The Audubon Zoo's patrons could eat inside or under a covered patio with a view of the newly landscaped grounds. Thirsty Prospect Park Zoo visitors could enjoy a beer from the new eating facility. The Central Park Zoo restaurant was decorated with murals painted by WPA artists. The Brookfield Zoo restaurant included a sculptural screen, wall decorations, and 36 "really beautiful" tables designed by WPA artist Frank Winters. Although Winters' services cost the Zoological Society nothing, the materials — state-of-the-art inlaid Formica — averaged $35 per table.[120] That the society decided to spend this extra money in an era when each budget item was scrutinized closely reveals that it sought to give Chicagoans of all incomes a first-class experience.

Leisure, entertainment, and enlightenment. The New Deal zoo promised to deliver these for low or no cost. Beautiful architecture, artistic adornments, and upscale amenities such as full-service restaurants meant that all Americans had access to some of the finer things in life. For financially strapped Americans, and even for those who were comfortable, the zoo was becoming an increasingly attractive public institution. For the unemployed, zoo projects represented a paycheck. Politicians, recognizing that zoos' appeal cut across class lines and served multiple interests, were enthusiastic supporters. But the color line was not so easily crossed in the 1930s, and not all Americans equally enjoyed the new zoos.

African Americans and Zoos: Not Quite Equal

In theory, African Americans benefited from New Deal zoos, gaining access to free recreation and to employment through federal programs. Urban blacks particularly could take advantage of the close proximity of these zoos. And, unlike other public recreational facilities, it appears that zoos were not segregated; there were no separate zoos for blacks, and blacks were not denied admission to those used by whites. In some areas of the country, zoo directors presented this integration as normal or even desirable. In Seattle, where blacks were only one percent of the city's population and were barred from many restaurants and public entertainment venues, the zoo director noted approvingly that black children under the supervision of social welfare workers had visited in August of 1933.[121] Among groups holding picnic days at the zoo in July of 1934, the Negro Baptist Church and the Negro Harmony Art Club were listed.[122] Nonetheless, racial insensitivity and prejudice were strong in the 1930s, and in one report the same director both boasted that his zoo attracted "people from all walks of life and all colors" and described an incident involving one of the zoo's monkeys, whose name was "Nigger."[123] Moreover, segregation of public facilities was the norm in the Southern states. Thus, zoos were not oases of racial harmony. Indeed, under the surface of the federal relief agencies' propaganda about expanded recreational opportunities for all Americans existed the reality that blacks continued to experience discrimination of various kinds in New Deal zoos.

Dallas, like other Southern cities, had a segregated park system. It built separate parks, playgrounds, swimming pools, and golf courses for its black residents.[124] Some sense of the inequities in the "separate but equal" system can be gauged by the fact that in 1941 the city park system included 35 swimming pools, only one of which was designated as a "negro pool."[125] When the zoo was improved in 1933, a "comfort station" for blacks was added.[126] This detail reminds us that even though the zoo had been technically open to all, the reality of "whites only" restrooms and the absence of facilities for blacks had meant that black visitors' leisure was significantly compromised. Without a public restroom, black zoo goers could hardly enjoy a leisurely day at the zoo. But, even after the construction of black-only facilities, blacks gained only a "separate but equal" status.

Minstrel shows were popular in the Southern states, where they were often performed for free in public parks — sometimes even by park departments and zoo employees. A *Dallas Times-Herald* writer organized "Municipal Minstrels" in 1922 to raise money for the purchase of a zoo elephant, and the Dallas Zoo Commission's meeting minutes of 1930 listed the recognition of the "Honorary Life Members of Zoo Minstrel Participants."[127]

African Americans in New Orleans surely felt uncomfortable in that city's parks, where white citizens complained if the city did not organize a sufficient number of minstrel shows. The elite who controlled the Audubon Park Zoo had to respond to citizens' expectations for free entertainment, including minstrel shows. Mayor Robert Maestri received a letter from Edward O'Neill, a citizen who demanded "dance schools, minstrel troupes, and vaudeville groups along the lines they had in the City Park." He reminded Maestri that "as a brat" [the mayor] had enjoyed "many a concert — movie, vaudeville show in the City Park." O'Neill was convinced that plenty of people would attend a city park concert-minstrel show to be offered by the apparently popular "BKA minstrel troupe," and he could not understand why Audubon Park did not offer similar entertainment. "What the heck's the matter with the present Audubon board?" O'Neill wanted to know. Had they "forgotten that they were kids themselves once upon a time? And not always wealthy men and millionaires? Some of them I know, as children did enjoy movies and vaudeville in public parks, including Audubon park. Of course, they, under their God blessed circumstances, personally care nothing for free music-movies-vaudeville public parks shows, but what of their unfortunate fellow men of New Orleans who cannot spare the dime to a quarter necessary to pay for their and their children's movie tickets? Are they to stay home and twiddle their thumbs? THEY have no well-filled coffers, no limousines, no country homes or summer palaces to while away THEIR leisure hours. O well Mr. Mayor, happy days will soon be here again, let's hope in Audubon park."[128] Here was the underside of a class-based approach to free recreation.

Although white citizens in Louisiana wanted to watch minstrel groups play, they were not as eager share the park or the zoo with real African Americans. The Audubon Park Commission spoke eloquently about parks as "breathing places, where children may romp in healthy activity and the weary toiler may breathe the balmy air of wood and pasture," but black children and toilers were not made welcome in its park.[129] Jim Crow laws in New Orleans segregated African Americans throughout the city. Blacks, for example, could not use whites-only branches of the public library system. The Audubon Park, site of the city's zoo, was not a whites-only facility, but it had different rules for whites and blacks. In 1939, park employees began enforcing a rule to keep blacks "moving all the time and not be permitted to use benches."[130] In other words, African Americans were allowed to walk through Audubon Park, but not "use" it. The park's history — it had been part of the Foucher and Bore plantations — had a firm grip on its present.[131]

The Audubon Park Board's order to keep blacks moving was not enough to satisfy all white residents. In January 1940, Mayor Maestri received a letter from a local women's business organization that protested African Americans in the park. It read, "Well, please do something about this — and it better be soon — we are a strong organization and it won't take much to put you on the 'spot.' We are watching you closely, and expect to see a change soon — we don't want our children rubbing elbows with Niggers. Better do something about this at once, or we're liable to appeal to the Newspapers — and this will cause you much embarrassment — we might even force you to resign — so get to work and put a stop to this."[132]

In another example, Park Superintendent Neelis received an anonymous letter that asked, "since when has Audubon Park become a hang-out for Negroes? Up until a few months ago a Negro was rarely seen in Audubon Park unless accompanied by a white child — but today, the park is really overrun with them — they are commencing to be a menace — when anything is told to them they give you a mouthful of sass. Why a white man really hasn't a chance to see the animals unless a big buck Nigger is hanging over his shoulders. We are Southerners down here, and there is no reason why Negroes should be crammed down our throats — they have their own schools — their own playgrounds — and they even sit in different parts of the streetcars and shows from us — why should they be permitted to roam around in our park with our white children. Audubon Park is a recreation for the white man — not the *Negro*; it is the white man paying the taxes for the upkeep of the park." The author noted that "upon inquiry, we were informed that you had given them permission to visit the Zoo because it was educational — a very poor excuse — very, very poor. Since when does visiting the Zoo imply knowledge — we don't go to the Zoo to get knowledge — we go there for entertainment — but of course if you insist on giving the Nigger his education by allowing him to visit the Zoo, then why not set aside one day out of the week for such purposes — say a Monday when things out there are slow — from Wednesday on the Park is full of white people, and they are really commencing to pass remarks of dissatisfaction regarding the liberties of Negroes in Audubon Park, for they not only visit the Zoo, but they are going further. They are getting bolder and taking advantage to (sic) the privilege allowed them and making use of the playground equipment thus preventing white children using the swings etc. Not only that but they are commencing to occupy the benches — in no time if this is not stopped, Audubon Park will be a total wreck. This is really going too far. They come to the Park in droves, driving through in their automobiles."[133]

In response to this hostility, the New Orleans branch of the National Association for the Advancement of Colored People (NAACP) appealed to Maestri to find ways to make African Americans feel that the park belonged to them as well. Louis Blanchet, the secretary in 1940, for example, asked the mayor whether there might be "some steps" he could take where "colored New Orleans may enjoy some part of the Audubon Park ... for some of its recreational activities." He noted that there were "many citizens who would like to frequent these parks and feel free to do so without feeling that they are violating any written or unwritten laws, or being subjected to reactions without proper protection from the park officials and city officials." There were "no places to which colored New Orleans may now go unhampered or free from liabilities they are not able to protect." He felt that "a consideration from [the mayor's] office would help the health, the morale, and the good feeling between the people that are now living here and who are here to stay. A contented people are good citizens."[134] Blanchet described a situation where legal segregation and social norms ("unwritten laws") combined to make the park and its zoo off-limits to New Orleans' African Americans. Worse, Blanchet's willingness to accept a plan that would enable blacks to enjoy even "some" of the park's amenities reveals the virulent grip of racism. The national reports from FERA, the CWA, and the WPA could speak eloquently about equality and opportunity, but on the ground in places like Louisiana, the rhetoric did not match the reality.

Just as New Deal zoos were technically open to all, so too the New Deal relief programs theoretically provided employment opportunities to African Americans. Black voters and leaders heavily supported FDR in the 1936 election, and they appealed to him to ensure that federal programs dispersed benefits in an even-handed manner. [135] The inclusion of African Americans in the New Deal's various relief programs was a major factor in black voters' shift from the Republican to the Democratic Party, and the evidence shows that black laborers shared in the construction of America's zoos, even in the South. Still, the black experience was not completely equal.

To a certain extent, African Americans' demands for equality were met by the New Deal. The Works Progress Administration successfully forced the employment of African Americans. The statutory language of the Civilian Conservation Corps (CCC)—a program designed to put men to work on projects such as building dams and fighting wildfires—explicitly forbade racial discrimination by requiring that local officials administering the program employ their fair share of blacks. [136] Comprehensive evidence of racial equality on New Deal zoo projects is unavailable, but the record provides suggestive glimpses of both equity and discrimination. On the positive side, photographic evidence shows black and white laborers building Midwestern zoos. Indiana's Mesker Park Zoo construction utilized an integrated CCC unit. Black WPA laborers worked alongside whites on the Toledo Zoo. An African American supervised the painting of murals at the Pittsburgh Zoo. And, as we shall see, black laborers outnumbered white ones in the early stages of the WPA's Audubon Zoo project.

In general, however, the Southern experience did not escape the region's deeply engrained racism. Southern Democrats did not all embrace FDR's New Deal relief programs, and some white Southerners were decidedly unenthusiastic about the direct aid or work relief of African Americans. Plantation owners particularly feared that low-paid African Americans would desert farm work for more lucrative relief jobs.[137] CCC integration rules were simply ignored in the South. For example, counties in Georgia with a population of more than 60 percent African Americans employed few or none in the CCC.[138]

Nevertheless, more African Americans than whites were employed as laborers for the WPA projects at Audubon Park. For example, African American workers built the Batture Levee in December 1935. When they finished, they shifted their work to the zoo. Although the Park Board did not keep precise statistics on the racial composition of every project at the zoo, they did keep general figures. They reported that work was started on October 23, 1935, with "346 colored labor and 203 white labor, and has proceeded with an average of 400 colored to 220 white labor."[139] Although a greater number of African Americans worked on WPA projects in this context, they typically earned less because they were more likely than whites to be employed in lower-paying job categories. In 1933 the pay rate for a skilled worker or foreman in New Orleans was $1.00 per hour. In contrast, the pay for unskilled workers, where African American men were primarily found, was only 40 cents per hour.[140] Moreover, white and black work crews were segregated, with blacks working a morning shift from 6:30 to 11:30 A.M., and whites working an afternoon shift from noon to 5 P.M.[141]

After the Audubon Zoo opened, the presence of black relief workers engendered greater controversy. Just around the same time that white citizens were complaining about the presence of blacks in "their" zoo, African American youth found employment through the National Youth Administration. In 1941, 120 "NYA colored boys" began work in Audubon Park, with many painting cages, rails, buildings, and concession stands in the zoo.[142] This was a short-lived experiment, and these workers were "ordered taken from the park" a few months later, with their employment regarded as "a loss to the park."[143]

White zoo visitors were quite prepared to view blacks as different, even in the absence of overt racial hostility or segregation. The fact that many of the zoo's most popular creatures came from Africa led to numerous associations between African people and exotic animals. A heavily advertised feature of the Ringling Brothers and Barnum and Bailey Circus's 1930 season was its "tribe of genuine Ubangi savages." Described as having "mouths and lips as large as those of full-grown crocodiles!," the Ubangi tribesmen and women were, in the words of one circus man, "a sensation." The sensationalistic circus poster both connected Africans with animals by comparing them to crocodiles, but also promised that the show was the "greatest educational feature of all time." Such a claim reminds us that "educational" was sometimes code for "sensational." As in the zoo, if visitors saw something new, the effect was "educational." Circuses played up the race dimension of their enterprise even when genuine Africans were not available. African-American canvasmen (those who assembled the tents) dressed up as "cannibal chieftains," complete with nose rings, and rode leopard-skin decorated elephants.[144]

Perhaps the most disturbing intersection of race and animals was located not in the zoo but in the stories told by the men who captured animals for the zoo. A recurring narrative theme, occasionally documented with a photograph, had African women nursing nonhuman primates. Explorer and animal dealer Julius R. Buck described the capture of the infant Bushman (a lowland gorilla eventually bought by Chicago's Lincoln Park Zoo) with the help of 17 men of the "Batwe tribe of little people."[145] Because Bushman was so young, Buck "secured a woman to nurse him through" a year-long stay at the American Presbyterian Mission in Yaounde (Cameroon). Elsewhere this woman is described as a "wet nurse," but Buck is vague on this point, evidently feeling that the details of keeping a baby gorilla alive for this length of time were not as interesting as other facts about the animal's journey to Chicago. He does say that the Africans spent long hours feeding the gorilla and keeping it entertained to stave off boredom. Buck notes, for example, that later he hired a "Native boy to care for and amuse Bushman."[146] Similarly, the famous animal trader Ellis Stanley Joseph reportedly saved an infant chimpanzee in Sierra Leone by "persuading a native woman to nurse it for several weeks."[147] A revealing picture from another gorilla hunt provides graphic evidence that African women literally served as wet nurses for the gorillas. A gorilla hunter named Philip [last name illegible] sent Marlin Perkins a photograph of another future member of the Lincoln Park Zoo appearing to nurse at the breast of an unhappy looking young African woman.[148] Although sanitized stories about animal captures like Bushman's made it into the newspapers, this photograph did not. Zoo directors apparently realized that there were publicly acceptable definitions of nurse and unacceptable ones as well. Within a small circle of directors and animal collectors, African American women could serve as wet nurses for animals, while white wives at home could raise the animals to be civilized.

Researchers of the era sought scientific evidence to support the racist assumption that blacks were "closer" to non-human primates than were whites. Harold Bingham, who conducted research at the San Diego Zoo, was engrossed by studying the sexual behavior and physiology of primates as a way (theoretically) of learning about human sexuality and evolution. In particular, he wanted to explore the physiological similarities between humans and gorillas because they suggested that humans were closely related to these animals and therefore Darwin was correct. In contrast, a Munich anatomist claimed that humans could not be related to gorillas because their genitalia were so different. To settle this debate, researchers examined human female fetuses for their genital resemblance to chimpanzees, and these fetuses were selected partly on the basis of race. A scientist named Chapman, for example, examined "two negro fetuses about five months old" at the same time that he was dissecting a female Chimpanzee.[149] From this exercise, Chapman concluded that the Chimpanzee and human female genital forms were quite similar. It is in the context of such racially charged research that we should understand the practice of African wet-nurses for infant, non-human primates.

While newspaper coverage of zoos may have been mostly free of overt racial animosity, racial stereotypes could play themselves out in stories about the animals themselves. Zoos gave animals names that were racially charged, and tongue-in-cheek reports about animal antics sometimes derived much of their humor from allusion to race. Stories about monkeys, particularly, could easily take on racial dimensions. Seattle's Monkey Island, constructed by the WPA and opened in 1940, was a hit with the local newspapers because of the constant struggles for dominance among the island's residents. The leading actors were given names — Chief Seattle, Queen Coco, Peter the Gook, Ting-a-ling Sing — that enabled reporters to frame the monkey's antics as political struggles among various ethnicities and between the sexes. One paper editorialized that these struggles were natural, since getting along with "other families with characteristics differing radically from ours" was not easy. The monkeys' fighting was funny because "they are so much like humans."[150]

This resemblance to humans had figured prominently even before the construction of Monkey Island, and it is clear that race news sometimes drove zoo coverage. Such was the case when a Seattle reporter wrote a lengthy piece about a fight between two monkeys at the Woodland Park Zoo, casting the episode as a boxing match between a "black" and "the great white hope." The article is a parody of Joe Louis's July 4, 1934, bout with Jack Kracken, except that in this case the white challenger is victorious. "The belief that fighters cannot come back, so common since July 4, was thoroughly and effectually disproved this morning when Chief Seattle, ex-champion of the Woodland Park monkey cage came again into his own to the immeasurable sorrow of Eureka who only yesterday proudly bore the laurels." The racial dimensions of the fight were explicit: "the color element entered into the fight to swell the general interest. Eureka is a black-faced monkey from Central America and Chief Seattle, native of Japan, played the Caucasian." The bout began after zoo Director, Gus Knudson entered the cage and Eureka snatched a cigar from his pocket. This theft "raised the old spirit of fight" in the former champion, Chief Seattle, whose previous defeat had seemingly extinguished "the hope of the white race." Springing quickly at the "abysmal brute" Eureka, his white challenger disposed of the "dark-hued scrapper" in two rounds.[151]

This story reminds us that the Depression-era zoo was not free of racial tension. However much boosters spoke of providing recreation and education to the people, their egalitarianism had racial limits. Free zoos and their amenities were not equally enjoyed by all citizens. The assignment of African Americans in zoo projects to laboring positions reflected the larger tendency of New Deal policy makers to allow inequality to exist within New Deal programs.

Visitors in Need of Discipline

For all of the official rhetoric about the wholesome benefits of public parks, in private, park professionals and politicians expressed considerable concern about the pervasiveness of criminal and antisocial behavior in these public spaces. At the Depression's onset, New Orleans' Audubon Park was "notorious as a hangout for hoodlums, dope fiends, and perverts," and even after seven years of New Deal improvements, the park had not improved significantly. In one 30-day period, toilets were vandalized, 25 lights broken, and a 14-year-old girl was assaulted.[152] After the Brookfield Zoo's opening weekend, park guards collected three bushels of empty wallets discarded by pickpockets.[153] Patrons were not always well behaved toward the very exotic animals they professed to enjoy seeing. Huge crowds meant a proportion of unruly visitors, and zoos responded with various mechanisms to control and shape the public's behavior.

Zoos and parks were both solutions to juvenile delinquency and the victims of it. Zoo directors and parks advocates claimed that zoos and playgrounds prevented juvenile delinquency by giving children something productive to do with their time. The recreation movement advocates claimed not only that outdoor play promoted children's health, but also that it protected children from accidental injury. Seattle's park department took credit for a three-fold reduction in fatal street accidents involving juveniles, and a city study demonstrated that rates of juvenile delinquency were proportionally lower in districts with generous playground and park facilities. Everett Smith, a Juvenile Court judge, theorized that in parks and playgrounds boys could "blow off" their "repressed energies" and "surplus emotions" that would otherwise find destructive outlets.[154] Seattle's director, Gus Knudson, enthused that "while children are at the zoo enjoying [animals'] antics, they are off the streets, thereby being saved perhaps, from a criminal career."[155] This view was taken on faith by those in the parks and recreation movement, who argued that "adequate recreational facilities" were an effective "means of curbing juvenile delinquency."[156]

Nonetheless, zoo directors and keepers had many problems with delinquent behavior, including profanity and attempts to harm the animals.[157]

Because zoos charged no admission fee, children used them like parks, entering and exiting as they pleased, unaccompanied by adults. Many boys, it seemed, blew off their excess energy by exercising their repressed hunting instincts. At the Seattle zoo, for example, small boys climbed the elephant's fence and poked the animal with umbrellas. They put "burrs and fishhooks in the monkey cage." They got near enough to the baby camel to pull out its hair. "They fed terrible things to the black bears and bothered them when they slept."[158] They brought BB guns and slingshots to zoos, engaging in target practice on the animals. In the absence of modern weapons, bottles and rocks were utilized.[159] Such unsupervised behavior led Robert Moses to support a new playground adjacent to the Bronx Zoo on the grounds that "it is of just as much advantage to the Zoo as to the Park Department to keep these children out of the Zoo."[160]

To combat this violence, Seattle's Woodland Park Zoo hired a teenager named Joseph Hahn to prevent patrons from injuring the animals. His boss, Gus Knudson, gave him a club and permission to use it. Within a short period of time Hahn had made 54 arrests.[161] Throughout his directorship, Knudson complained about the property destruction caused by "vandals" and "hoodlums," and many of the repairs he reported each year were designed to make facilities vandalism-proof by, for example, covering glass windows with metal screens. In 1935 Knudson observed that "people are getting very destructive about the park."[162] Public restrooms were a problem, and many zoos employed attendants and even charged an admission fee to "keep the places clean and guard against misconduct and vandalism."[163] Even signs themselves were subject to vandalism, liberated by souvenir seekers.[164] Such behavior was rampant in Seattle's city-wide park system, exacerbated in part by budget cuts which eliminated the park police system in 1931. The Park Board appealed to the police department to help address the "considerable difficulty with boys on our playgrounds."[165] As a stop-gap measure, Seattle gave select park employees special police badges. Gus Knudson applied for and received such a badge in 1936.[166]

Adults too required disciplining, and it is evident that middle-class norms for appropriate behavior in public places had not yet taken hold by the 1930s. When Knudson returned in 1930 from an investigation of 26 zoos across the nation, he reported "a great many restrictions on the public in all the parks" imposed through signs of various sorts. Some of these signs — "no feeding of animals," "no smoking in buildings," and "no littering" — would be familiar to today's visitors. The extent to which these prohibitions were ignored was shocking, however. It took three Woodland Park Zoo employees working all day Monday to clean up the garbage "strewn around by picnicking citizens on Sunday."[167] The situation was just as bad in New York City, where the amount of garbage deposited around the Bronx Zoo's buildings, walks, and lawns was "quite unbelievable until seen." Sunday's refuse, much of it dropped within sight of one of the 75 garbage cans, took a "small force" one and one-half days to clean up. To combat this "disorderly" and "objectionable practice," the director instituted a campaign for "cleanliness and decency," authorizing zoo employees to issue a summons to all offenders.[168]

Other signs reveal that visitors made little distinction between a park and a zoo: "no ball playing on walks or lawns, do not cut or deface trees or railings, no kite flying, keep to the right," and many more. Athletic activities on zoo grounds were prohibited. These regulations were "strictly enforced" by guards or police who reported to the zoo director.[169] Although sometimes park-like, zoos were not good places for park activities. Conflating the two could be dangerous to animals and visitors alike. In 1919, a young girl playing at the Bronx zoo was seriously mauled by a bear when she crossed a low railing to retrieve a ball that had rolled near the bears' cage.

Although children instigated their fair share of zoo mayhem, their elders were often blamed for anti-animal behavior. When a trained seal at the California Academy of Science (San Francisco) died after ingesting four pounds of coins, a zoo newsletter editorialized that "it is a ten-to-one bet that you will find an adult visitor, and not a child, at fault."[170] All zoos struggled with visitors who fed the animals or, worse, threw inedible objects into their cages. Many an animal faced a daily "barrage of peanuts" intended to stimulate action of some kind.[171] Autopsies frequently revealed that an animal died because of accumulated glass, rubber, and so forth in its stomach. In response to these dangers, zoos posted "do not feed" signs, provided appropriate foods for visitors to buy, and even pressed for the passage of criminal codes regarding feeding. New Orleans made it illegal to feed or give any object to any animal in the Audubon Zoo, with penalties of up to a $25 fine or 30 days in jail.[172] The New York City Parks Department posted signs to inform visitors that feeding the animals was a $25 offense. One young boy discovered the penalty could be much greater: hopping the guard rail around the Sun bear cage, he held out some peanuts for the bear, who took the peanuts and then tried for his forearm as well.[173]

Teasing of animals could also be a problem, and not just for the teased. In one example, Bondo, a San Diego Zoo chimpanzee, got a kind of playful revenge on some visitors tormenting him in his cage. Five sailors were standing around his cage and poking a piece of galvanized tire wire into the enclosure. Bondo reached for the wire as it slid into his space only to find it quickly pulled out of his grasp. Sadly, the wire sliced his hands as it slid back out. Bondo, however, quickly figured out the game, and the next time the wire came in, he signaled to a fellow chimpanzee sitting high up on a ledge who spat a mouthful of water on the sailors while Bondo grabbed the wire. This time the wire cut the hands of the sailors who were both embarrassed at being spat upon by a chimpanzee and now had injured hands as well. Although Bondo wanted to continue playing, the sailors refused. Interestingly, zoo Director Belle Benchley watched this whole episode but declined to intervene. She silently applauded Bondo's winning the game, but for reasons that are unclear, she had faith enough in the visitors that they would not seriously hurt the chimpanzees by putting wire into their cages.[174] In other cities, teasing — at least after hours — was a crime and could land the offender in jail.[175]

Visitor cruelty to animals could become quite extreme. Stone-throwing visitors killed a sea lion at the National Zoo, and incidents of this nature were repeated across the country with ghastly frequency.[176] In what appeared to be an early drive-by shooting, unknown perpetrators cruising the Dallas Zoo access road killed two deer and a zebra with rifle fire. Another vandal poisoned Doug, the zoo's young chimpanzee.[177] Poisoning occurred with disturbing regularity in America's zoos. An autopsy of an Audubon Zoo seal revealed the cause of death as "malicious poisoning."[178] Seattle's Woodland Park Zoo lost five cougars to deliberate poisoning in 1936.[179] Animals in the St. Louis Zoo also fell victim to poison: "The second poisoning within a year took place Sunday at the St. Louis Zoo, it was announced yesterday by Director George Vierheller. The victim was Bobby, a 15 year old organ-outang (sic) valued at $2500, which died within a few minutes after strychnine had been tossed into his cage in the Primate House by an unidentified person." The previous summer six mammals had died from strychnine poisoning, but Bobby's death was particularly galling because the killer had apparently acted under the very nose of Vierheller while he and other employees were filming an animal performance.[180]

Newspapers editorialized against such behavior, seeking to make it socially reprehensible. A writer for the *St. Louis Star* opined that "It's hard for the average human to discuss the poisoning of 'Bobby,' the 140 pound performing orang at the zoo, in printable language. The dog poisoner is a familiar type of degenerate. The impulse that leads some so-called humans to lay out poisoned meat, or to throw poison into the cages of any helpless animal, can't be set down altogether as criminality. There must be an actual streak of insanity in it. When it comes to causing the agonized death of an animal nearest human in its intelligence and affections, as in the case of this young orang, the most charitable comment is that the perpetrator is a subject for an asylum and the insanity experts." The paper recommended hard labor for the culprit, who had "robbed the public." Ideally, however, the punishment ought to be more fitting: "One cure for it might be to put the offender, when caught, in a cage and leave it to the public to say what he gets to eat."[181]

Putting offenders in cages was not a serious option, but zoos could protect animals through cage design. In a physical design sense, zoos controlled public behavior by limiting access to the animals. The earliest public menagerie was often just a collection of cages, small buildings, and enclosures scattered in a haphazard pattern through a portion of an existing park. Such menageries might not only lack perimeter fences, but might also allow visitors to drive their cars right up to the animals. Like the parks, these unfenced zoos were effectively "open" 24 hours a day. In Chicago, people made homeless by the Depression slept in the Lincoln Park Zoo.[182] Risks to animals and the public were potentially serious. Escaped animals easily made their way into surrounding neighborhoods, while vandals and dogs had unsupervised access to zoo grounds.[183] Gus Knudson returned from his zoo tour with approving comments about the large number of zoos that had fenced perimeters and clearly established opening and closing times.[184] In 1933, when CWA workers became available in Seattle, Knudson requested that they build a fence around the entire zoo, as well as an "entrance arch."[185] Instead, "against [his] wishes," workers were directed by the park commissioner to remove one section of fence, allowing "the public" to enter the park "from all direction."[186] Zoos in Toledo, Dallas, New York City, and elsewhere first received perimeter fencing and entrance gates thanks to New Deal programs, although others, such as the Audubon Zoo, remained unfenced despite appeals from zoo staff. Another perimeter improvement was the entrance gate, which was a particularly symbolic way to order the flow of people into the zoo, rationalizing the space. Instead of entering from any convenient opening, visitors filed through clearly marked gates. Vandals, unfortunately, were not deterred by fences, and the Bronx Zoo undertook "strenuous and constant" efforts to prevent the rampant vandalism and theft inflicted by after-hours fence climbers.[187]

Inside the New Deal zoo, the public encountered formal walkways and directional signs, further guiding and limiting public behavior. Prior to New Deal improvements, the zoo experience could be relatively laissez-faire, with visitors wandering on unimproved and ad-hoc paths from one unmarked exhibit to another. With New Deal funding, however, directors could build permanent brick or concrete paths. Tapping the skills of Arts Project woodcarvers, zoos gave visitors directional and informational signs. Moreover, WPA artists and writers produced zoo maps and guides that further rationalized the zoo-going experience. The maps — with their emphasis on paths and destinations — not only disabused visitors of the impression that the zoo was an open space that happened to house animals, but also provided visual models of appropriate zoo behavior. Map borders contained drawings of children and families engaging in polite and acceptable behaviors.

Ironically, the desire to provide the masses with a first-class experience put new consumption pressures on poor Americans. The formalized zoo space often included a feature old zoos lacked — a restaurant. Instead of bringing a picnic basket to the zoo, citizens were now subtly encouraged to buy their lunch at the zoo restaurant, which could charge inflated prices because it had no competition. Class distinctions became evident, as those who could afford the zoo's food enjoyed pleasant indoor or outdoor seating in beautifully designed settings. The zoo restaurant thus established a middle-class norm for zoo visitors.

In the various justifications for public zoos were implicit assumptions about the social and economic class of typical visitors. Although Americans of the "better" classes enjoyed zoos, they were obviously only a small percentage of the hundreds of thousands or even millions that visited a big-city zoo each year. Most zoos charged no admission fee because they correctly assumed that such a fee would be either a barrier to many Americans or an unjustified burden to them. One well-to-do citizen complained about the high prices at the Bronx Zoo's cafeteria "on behalf" of visitors "in obviously poor circumstances." In his view, the zoo should be free all of the time (some days required tickets) and its concession prices should be affordable to all. In the big scheme of things he felt that any revenue the zoo might gain from admissions fees and concessions "is not sufficient to serve nearly as important a purpose as a free park and zoo for poor people, whose ability to find recreation within their means is very limited."[188] The existence of "pay days," however, suggested that many of those people from the better classes did not want to rub shoulders with their poorer brethren.

Even a zoo's public facilities were not necessarily accessible to all visitors. The National Zoo charged no admission fee, but its restaurant — built with New Deal funds in 1940 — catered to Washington's elite businessmen and government officials. Managed by Gordon Leech, the restaurant specialized in four-course game dinners, serving venison, bear, pheasant, whale, iguana, and even elephant to members of an unofficial "Anteaters Association." Although alcohol was not allowed on zoo grounds, Leech obtained a license to serve beer and wine to these club members. The special game meals were quite expensive and were beyond the means of the average visitor.[189]

Although zoo directors and their allies in the press skillfully argued that zoos should receive funding from the New Deal because these were institutions for deserving people, the reality within the border of the zoo was more complex. Class and racial privileges could be had for those willing to ask or for those whose patronage zoo directors sought. Sol Stephan, the Cincinnati Zoo's director, conducted private, after-hours, behind-the-scenes tours for well-connected patrons. Marcia Davenport, a member of the Luce publications family, appealed directly to Robert Moses for permission to see the Bronx Zoo's three new baby tiger cubs, which were being cared for by the tiger keeper's wife in her home. Assuming that the keeper's wife could adjust her schedule, Mrs. Davenport even gave the time and date on which she wished to visit. Moses obligingly conveyed the request to the zoo's director, with the none-too-subtle hint that it should be honored.[190]

In significant ways, then, zoos were built by and for the masses to amuse, educate, and serve as a controlled space that hopefully offered alternatives to less desirable pursuits while at the same time bringing in revenue for surrounding business. Supportive newspaper editors and publishers recast the debate about New Deal funding away from a discussion about whether it was better to use New Deal money and local taxes to help starving people or let exotic animals in the city's charge die, to an easier contest between deserving Americans and uncaring politicians stealing money from a worthwhile recreational and at times educational activity. Perhaps most surprisingly, reptiles were one of the central kinds of animals that zoos used to draw Americans through their gates during this period.

Four

Why Snakes? The Spectacle and Science of Snakes

It is amazing how many people are attracted to snakes.[1]
—Gus Knudson, Woodland Park
Zoo Director, 1934

The reptile house emerged during the 1930s as a particularly popular attraction at the zoo. By the early 1940s, the Bronx, National, Staten Island, Brookfield, Cincinnati, San Diego, St. Louis, Toledo, San Antonio, Buffalo, and Philadelphia zoos had large, modern "reptile houses" (many built or improved by the New Deal) managed by curators who were trained herpetologists, sometimes published in the field. Indeed, newspaper reporters frequently relied on reptile curators as the top authorities on these animals. The fact that reptile keepers were almost always referred to as *curators* indicated that their scientific knowledge put them on par with reptile experts at other scientific institutions such as natural history museums. Reporter William Gaines informed the citizens of Alaska that "Raymond L. Ditmars, curator of mammals and reptiles at the [Bronx] zoological park, knows just as much about ... reptiles as any man alive." An unnamed writer for Ogden City's (Utah) *Examiner* told his readership that Ditmars was "just about the world's foremost authority on snakes."[2] Reptile curators at large American zoos during the 1920s and 1930s collected and displayed snakes in a highly scientific manner, and they supported field and laboratory research. A typical New Deal reptile house was state-of-the-art. It took advantage of current knowledge about cage design and temperature control to give each snake species an enclosure that promoted its health. Specimens were displayed behind glass

Ground Floor Plan, New Zoological Building

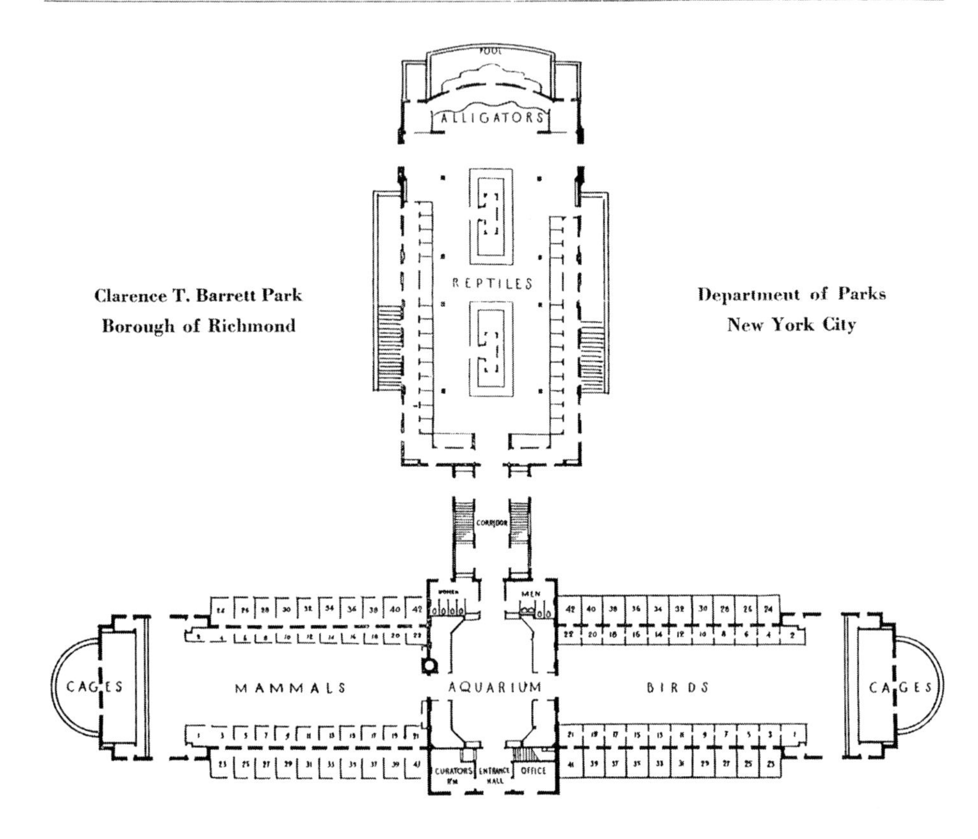

Ground floor plan for the Staten Island Zoo (1935). Notice the relatively large amount of space devoted to the reptile wing, a common feature of New Deal zoos (Staten Island Zoological Society).

and arranged systematically, by species or geography. WPA artists painted background panels to depict each snake's natural habitat.

But should snakes get any money at all? They were not the fuzzy, feel-good animals that one might have imagined would have received taxpayer money at a time when many Americans were going hungry. This was a question asked by Mrs. Charles J. Dolan in her editorial to a St. Louis newspaper: "What are the St. Louisans thinking of to let the gentlemen of the Zoo Board go ahead without protest with their amazing proposal to squander $100,000 of the taxpayers' money on a climax of absurdities — and worse — a chamber of horrors that nobody wants — a snake house." In his coverage of keepers force-feeding a python at the St. Louis zoo, a reporter for the *San*

Antonio Light noted that "in these days when so many human beings are going hungry, the spectacle of the big snake turning up its nose at food such as a spring chicken is a strange one."[3] Even some politicians found little to like about reptile houses. When former New York Governor Alfred E. Smith spoke at the Staten Island Zoo's opening ceremonies, he likely expressed the opinion of many New Yorkers when he admitted, "I'm not particularly attracted to snakes" and expressed great admiration for St. Patrick because he "banished them out of Ireland." Smith captured a common sentiment that zoos needed warm and fuzzy animals, not small and dangerous ones, when he declared that the zoo needed some lions. Just days after the zoo's opening ceremony, the *Staten Island Advance* editorialized, "Bring on Those Lions." While the snakes were "interesting," they "leave us cold."[4] The newspaper ran a cartoon showing a man cowering behind a tree as he peered fearfully at the Staten Island Zoo's new reptile wing.

Snakes were largely fearsome, repellent creatures in the public imagination, yet, paradoxically, they were irresistible. A newspaper in Seattle reported, "Seattleites want snakes." One woman at the Woodland Park Zoo commented that "the snake house was the first and last place" she visited on each trip. Some visitors, initially reluctant to even enter the reptile house, soon made it their primary destination. One family got a huge "thrill" from the snakes, "the first they had ever seen." Reptile houses were added to many zoos during the 1930s, no doubt in part because "snakes are very inexpensive to keep and are attractive to many people."[5] But these buildings were also important to men like Bill Mann and Raymond Ditmars who perceived that they could support both their educational and scientific interests. Unlike the charismatic megafauna such as giraffes or elephants, reptiles and amphibians came in a huge variety requiring field classification and field study around the globe. And, Americans harbored many false beliefs about snakes, so the educational possibilities were rich.

The public's fascination with snakes encouraged zoo directors to build up their snake collections, but it also enabled zoo herpetologists to pursue research and publishing agendas that ultimately benefited both snakes and people. Zoo directors liked to justify their institutions on the grounds that they were "educational," and reptile curators self-consciously sought to expand Americans' knowledge about snakes. They wanted to help Americans avoid snakebites, but also to prevent people from slaughtering snakes needlessly.

Snakes in the News

Reptile houses and new snakes at the zoo attracted significant media coverage. News of the National Zoo's construction of its reptile house was covered in papers throughout the country. On August 23, 1934, the *New York*

Times reported on the arrival of 15 exotic snakes from the Malay Peninsula collected for the Staten Island Zoo's new reptile building. Zoo Director Carol Stryker, accompanied by the American Museum of Natural History's Dr. C. H. Pope ("an authority on the reptiles of the Malayan jungles"), met the arriving freighter *Eurybates* at Staten Island's Pier 20 to inspect the collection, which included black and yellow mangrove snakes, as well as ten dead cobras, kraits, and vipers from India. Most interesting to the press, however, were four "flying snakes," a tree-gliding species "distinguished by their great beauty and their bad tempers." As if to prove the latter characteristic, one of the snakes quickly escaped its crate with a great leap, slid up Stryker's shirt sleeve, and bit him, creating obvious amusement for the reporter but no long-lasting effect on Stryker (the snake was only "mildly poisonous").[6]

The media covered snakes in part because of how venomous they were. A large collection of South American snakes featured "seven species of the 'dreaded genus Bothrops.'" The Staten Island Zoo was pleased with a shipment of snakes from A. St. Alban Smith that included ten sea snakes — considered among "the most deadly reptiles known to man" — as well two spitting cobras, and a king cobra. "One discharge of its venom would be sufficient to kill ten men," the reporter warned. By 1938, the Staten Island Zoo could proudly claim "twenty-eight species of poisonous reptiles," many of which reptile curator Carl Kauffeld (who had once been bitten by a timber rattlesnake) described as "exceedingly tempermental (sic)." Kauffeld, a serious herpetologist, appreciated the media interest in the zoo but deplored journalists' "flamboyant inaccuracy" and the "great glee" with which they reported on mishaps involving snakes. When he accidentally brushed a finger against an Indian cobra's fang while handling it, reporters ignored the facts to create a sensational narrative. The brushing became a "bite," and the snake was transformed into a more deadly-sounding King Cobra. Or sometimes reporters captured the sentiment shared by many in the public that venomous snakes should be killed. "The only good Cobra," one reporter stated, was a "dead cobra."[7] Such stories emphasized the personal risk that zoo workers took in handling these snakes and seemed to promise a vicarious sense of danger for zoo visitors.

Despite, or perhaps because of, the fear and repulsion some Americans felt around snakes, Ray Ditmars and other herpetologists recognized that they generated good publicity for zoos. At the very least, reptile curators could present themselves as unafraid of snakes, perhaps garnering some interest in new reptile houses in the process. When Ray Ditmars finally received a long desired Bushmaster for the zoo, "thousands of people rushed up to see him," attracted by his fatal capabilities: "he was only a baby, [but] he packed a lot of swift and terrible death." When Ditmars and St. Louis Zoo curators organized a snake hunt in the Ozarks in 1931, local reporters turned the scientific expedition into a macabre horror story. Ditmars was cast as the fearless American hunter, prepared to enter the snakes' habitat with only "pockets of string" and a "good used stick," employing the "simple cowboy trick of lassoing his catch." Ditmars played up the significance of the event, claiming that he and "all of the scientific boys in the country who know their snakes" would embark on an adventure "destined to go down in history as one of the big captures of the Mississippi." Lest the reporters not appreciate the potential danger, he promised to "confine" his own hunting to rattlesnakes, which "hang so thickly on the ledges of what are known as bear dens ... that they resemble a long black carpet."[8] The image of a string-wielding Ditmars entering these dens no doubt fascinated and horrified readers who perused the full-page, illustrated story of the hunt. For those who could not or would not venture into the wild to see these dangerous reptiles, a zoo's reptile house was a safe and convenient alternative.

The National Zoo's Snake House

In 1929, the Manns took their first married trip together — not for pleasure, but to tour zoos, with special attention paid to the design of reptile houses. Lucile, Bill's wife, quickly learned to empathize with the Munich Zoo director's wife, who joked that she traveled a great deal with her husband, but "all I ever see is a railroad station and the zoo!" Mann wanted the National Zoo's reptile house to compare favorably with the best in Europe. Accompanying the Manns was an architect, Mr. Harris, who was afraid of snakes. Bill believed that this was the best kind of architect to have for a reptile house because he would surely design exhibits that were "perfectly secure" against escape. Although they visited zoos in Budapest, Berlin, Hamburg, Cologne, Paris, and elsewhere, the London Zoo's reptile house most impressed Bill. The "presiding genius of the reptile house in London was Joan Proctor," who was "both a trained herpetologist and a competent artist." She designed the building and thus had to be "an architect, electrician, and mechanic in order to get what she wanted, but the building as it stands is a lasting memorial to the courage and the vision of the talented woman who planned it." Proctor was ill when they were there and died a year later. Prior to becoming curator at the London Zoo she had worked at the British Museum. Proctor had invented what Lucile called "gadgets" for the animals' "comfort and for the safety of the keepers working with them." Mann marveled at the experience of standing "in semi-darkness, in a long room with black walls and ceilings, and look[ing] into the brilliantly lighted cages of the reptiles, which line both sides of the building." The reptile cases were electrically wired for warmth, ensuring that the snakes would come out from hiding so that visitors could actually see them. As Lucile gazed at the snakes, their "coloring set off by the artistic backgrounds and natural surroundings," she was finally convinced of their "beauty."[9] But the engineering inventions were equally important for the animals because they led to greater health, movement, and longevity. The London Zoo was able to keep their reptiles alive, sometimes for years.

Mann's international travel and his professional involvement with the Herpetological and Vivarium Societies were politically important because they gave him credibility when he argued before Congress that a new reptile house would be of scientific value. He appealed to the American public for the building in a radio address on December 19, 1930. In the final draft he prepared before his talk he assured the public that it was a worthy use of tax dollars because it afforded Americans "the greater knowledge" of things scientific. By appealing directly to the constituents, he hoped they would urge Congress to furnish money for construction. He had asked Congress for money for the reptile house first because he thought it might be "hardest to get." He argued that "the word reptile has very little popular appeal, though it has a great deal of interest, and the idea of spending a large sum of money on a building in which to house snakes at first met with violent opposition." But his experience at other zoos made him confident that once the doors were opened it would be "the most crowded and popular house in the whole Zoo, as well as having the greatest educational possibilities." He believed that Americans were interested in snakes, but insisted that there "was a deeper ignorance about these animals than about any others." Even by that period Americans still believed tall tales about "the hoop snake, the cow-milking snake, the snake that breaks in pieces and then puts itself together again." More important, however, "each year human lives are lost through snake bite by using foolish remedies that have long ago been proved worthless instead of up-to-date simple and thoroughly efficient serums."[10]

Mann had ambitious plans for the Reptile House. In addition to housing reptiles, it would also exhibit insects and batrachians. He was, after all, "Mr. Ant," and he believed that the public needed as much education about these animals as they did about snakes. He was inspired by the insect collection at the famous Hagenbeck Zoo in Stelling, Germany, where there was a "remarkable collection of living insects." The entomologist in charge of them "took as much pride in his cockroaches, spiders and centipedes as a trainer took in his Bengal tigers." But the bulk of the space was for reptiles. By this time, scientists, including Mann, had identified "5,000 kinds of reptiles," "1,500 or more frogs and toads, and myriads of insects so the building had to be large." And it was. Nearly "200 feet long and 82 feet wide," it contained

"136 cages" ranging in length from "8 inch ones to hold tiny frogs" to one "enormous cage 80 feet long" that they called "crocodile land." This particular cage was "divided into three parts, each containing a lake and tropical beach, ideal homes for our large school of American alligators, South American Caymans, and American and Old World crocodiles." Mann and his predecessors at the National Zoo had slowly collected their animals through their own collecting trips, dealers, gifts, and trades from other zoos. They were waiting, for example, for 14.5 foot cobra, the "largest and most dangerous of all poisonous snakes; a "pair of green mambas from West Africa and a whole nest of rhinoceros vipers" from the forests of the same region; and a "friend" in Texas had "nineteen cages of rattlesnakes ready to send."[11] All they had to do was make sure that their innovative heating system worked because cold-blooded animals needed warm homes that were not available in Washington winters.

Mann had three related goals for the completed reptile house. The first was that he hoped that Americans would respect reptiles, but enjoy them as well. He felt that "familiarity should not breed contempt," which, he believed, would ultimately lead to the death of the person, rather than the animal. Instead, he hoped visitors would become "fond" of the animals. But the zoo was "first of all an exhibition," a kind of scientific circus, so he attempted to "have a representative collection of cold blooded creatures exhibited in the best possible way." Visitors would walk through the building and see the animals exhibited in their "natural habitat" and "come away from the building knowing the difference between a snake and a lizard," which "few people do by the way." This fit the mission of the institution because, he noted, the "National Zoological Park is a great public enterprise, established by Congress at the request of the Smithsonian Institution, for the advancement of science and the education and recreation of the people."[12]

The zoo's reptile house, in particular, was of national significance to herpetologists and the zoo community. The American Society of Ichthyologists and Herpetologists formally thanked the U.S. Congress for funding the building, which it called "an outstanding development in public education in America." In a show of appreciation, the society pledged to hold its 1932 annual meeting at the National Zoo. Mann's early experience with the construction of a reptile house made him a valuable resource to other zoos that built or planned reptile houses during the 1930s. He dispensed reptile house advice to Detroit, Cincinnati, and Chicago, for example.[13] The National Zoo's reptile house was one of the earliest of its quality in the United States, but within a decade similar structures were added to zoos in Toledo, Chicago, Cincinnati, San Diego, Buffalo, San Antonio, and Staten Island.

The new reptile house would require new specimens, so in 1930 Lucile accompanied Bill on a collecting trip to Central America, where Bill had extensive contacts among scientists doing work there. Technically, this was a "vacation," because it was not financed by the zoo. In Central America the Manns caught only a few snakes themselves, but they acquired a large number of them from an American in San Pedro Sula, Honduras, who extracted venom from poisonous snakes for antivenin laboratories. Lucile recalled that "his back yard contained cages of poisonous reptiles ready to be turned over to a snake farm, and we had obtained permission to buy for ourselves any of his snakes we wanted. As I remember, we wanted them all, and came away with something like twenty-one venomous snakes and a few that were harmless but rare." Transporting these snakes from San Pedro Sula to Tela (where they acquired additional snakes) required special assistance from Lucile. No live animals were allowed on the train, so she put the burlap bag full of poisonous snakes under her skirt so that no "curious native" would pinch the side of the bag and get bitten and so that the conductor would not see it as he passed by. Not all of the snakes made it back, however. In Honduras they had obtained "nine 'jumping vipers'—rare and poisonous specimens." These snakes were rumored to "spring into the air when they attack," and Lucile noted that "there is a record of a man on horseback being bitten above the knee by one." The Manns packed each snake species by itself because "most snakes are immune to their own poison." In this case, however, they realized that "seven of our nine jumping vipers were dead." They had attacked one another in the back. "Their bodies still showed the scars of battle, some had died in the act of biting and their fangs were still locked in each other's stiff bodies." They had "learned a lesson in natural history—the jumping viper is not immune to its own poison."[14]

By 1933, the heart of the Depression, Mann's nearly new Reptile House already needed repairs. There were cracks in the pillars of the north door, loose pillars at the south door, a rusted out door on the cobra cage, and a loose guardrail, among other hazards. Mann recognized that the Civil Works Administration (CWA) could fund this repair work, and he was adept at negotiating the bureaucracy to get maximum benefit for the zoo. He was not content, for example, to accept just any employees sent by the CWA. For the Reptile House repairs and other New Deal–funded construction at the zoo, Mann typically wrote to the CWA requesting specific employees. From Captain Howard F. Clark, the chairman of the Work Planning and Job Assignment Committee, Mann requested that he be sent "Mr. Smith Elsea," who had worked at the zoo "for some time past on the unemployed roll." Mann found him "thoroughly satisfactory" and hoped to have him reassigned to the zoo. Similarly, he wrote to David Ruml, at the U.S. Employment Service, hoping to re-hire John W. Chester. He was familiar with Chester's "work" and "character" because he had "worked in the park on various occasions." Mann sent Harry Humphrey to David Ruml's office at the U.S. Employment Service with a letter requesting Humphrey as an employee. As was typical, Mann noted that Humphrey had "worked in the Park before" and was "most satisfactory." He added that they would "appreciate it very much" if Humphrey were assigned to the zoo. In fact, Mann requested so many mechanical engineers that Captain Clark eventually phoned him and said that he could not send any more. Although engineers earned only $1.50 an hour, Mann was directed by the CWA to "get more work" out of the men already sent. President Roosevelt, however, wanted his legacy in stone, rather than on easily forgotten, but necessary grounds work. In 1933, for example, Roosevelt issued an order relating to the Public Works fund stating that "all men employed under the Civil Works project should be engaged in Capital improvement projects and not upon maintenance." So, CWA administrators suggested that Mann transfer all men funded by the Civil Works fund "from maintenance activities, such as assisting keepers," cleaning up the grounds and "other routine maintenance which leaves no lasting benefit" to working on buildings.[15]

Mann's decision to build a reptile house first was partly strategic, as at least one reporter recognized. In his column "On the Inside of Washington," Rodney Dutcher observed astutely that, despite William Mann's protestations that he needed to build a reptile house first because it was the least popular, nothing could have been further from the truth. In fact, he noted that Mann used the success of the reptile house to get Congress' approval to appropriate New Deal money for additional projects at the zoo. Some of the 2,500,000 visitors to the zoo who came to visit the reptile house had partly helped pay for additions to the institution through their federal tax contributions.[16]

Although millions of Americans visited the National Zoo during the Depression, educating people who went to the snake houses had its limitations. It left out all of those who, for one reason or another, had never made this particular trip. That left large numbers of Americans without the benefit of the scientific information that zoo directors believed was of life and death significance. To reach a larger audience, zoo directors and herpetology curators in zoos turned to the media to educate the American public about snakes in particular.

Educating the Public

For the popular press, herpetology curators wrote full-length articles about how to identify snakes. This was an extremely valuable public health service at a time when Americans did not have easy access to printed, illustrated guides of reptiles. (Although one of the most popular reptiles guides today, the Peterson Field Guide to *Reptiles and Amphibians of Eastern/Central North America*, was later written by one of these herpetologists.) Ditmars, for example, penned long articles helping Americans distinguish between venomous and non-venomous snakes. He wrote, moreover, in language accessible to Americans so that he could prevent bites. In one piece he told his readers that there were four kinds of poisonous snakes, but noted that the word "kinds" would not be "quite proper in a scientific description, but the present article is intended to be a plain statement, easy to remember." In another work he assured his audience that "the greater number of the snakes we meet on our hunts and hikes are harmless." He noted that the black, coachwhip, pine, red racer, "chicken," "milk," King, yellow gopher, and bull snakes were not dangerous. He advocated protecting these animals both for economic reasons — they ate rodents that plagued crop farmers — and for fun. He believed that "the protection of our animal life adds to the fascination of the outdoors. There's a kick in the thought that bears lurk in the mountains and a thrill in the thought of 'a snake in the grass.' Aside from the actual hunter, we all enjoy the woods a little more with the realization that varied life lurks around us."[17]

Ditmars' fears about the future of America's snake species were not misplaced. Agricultural and industrial development destroyed natural habitats, and unsympathetic Americans destroyed snakes themselves. Reptile house curators could thus claim that they were protecting some species by exhibiting them in the protected environment of the zoo. The Audubon Zoo used its acquisition of two rattlesnakes — *Crotalus adamanteus* and *Crotalus horridus*— to offer a lesson about why these "formerly abundant" species were now making "a last stand against civilization." Succumbing to the pressures of feral razorback hogs, drained swamps, and ever-larger human populations, these snakes were rarely seen anymore in Louisiana, and the zoo hoped to secure "records and specimens" of the state's various snake species on the "verge of extinction" — blue and brown bull snakes, and coach whips — "before they disappear entirely."[18] By specializing in its own state's snakes, the zoo would become an early conservation center of sorts for these endangered species.

Not all snakes were harmless, however, and Ditmars and other reptile curators wanted to protect Americans from bites. Although he described them in great detail, Ditmars did not recommend memorizing the appearance of the poisonous coral snake because he believed that people were easily confused when actually confronted by a snake. He pointed out that it was "a simple thing to get the wires crossed in a combination of this kind and pick up the wrong snake."[19]

Ditmars and other reptile curators recognized that America's snakebite problem might get worse as Americans flocked to the countryside thanks to cars and newly created roads; people would be bitten and snakes would be unnecessarily killed. Seeking to head off both eventualities, Ditmars and his fellow herpetologists sought out newspaper publicity. In a lengthy interview with a Missouri reporter, Ditmars described local snakes in great detail to help citizens identify which ones might cause them the most harm. And he offered careful advice about avoiding snakebites. He noted, for example, that hikers should avoid casually grabbing onto bushes or rocks to pull themselves up to higher elevations, noting that certain venomous snakes sometimes rested in these places and would likely bite the hikers when startled. Ditmars was particularly concerned about the increasing numbers of Americans who were visiting the wilderness by virtue of better access from cars, increased membership in scouting organizations, and greater recreational interest in the outdoors. He warned that "only caution can save the vacationist [sic] from snake-bite danger." He noted that the snakebite problem needed more attention because "the popularity of hiking, the unprecedented growth of the automobile camping custom of vacationers — all these feeders of the out-doors habit have exposed many new tens of thousands to the dangers of snake bite." In addition, he observed that the "conservation of forest areas and their systematic protection from fire means that Nature will have continually protected regions for the

Staten Island Zoo reptile curator Carl Kauffeld displaying the fangs of a diamond-back rattlesnake (1937). Like other reptile curators, he sought to educate Americans about snakes. Kauffeld wrote *Snakes and Their Ways* in 1937 (Staten Island Zoological Society).

propagation of such smaller wild creatures as are the natural prey of poisonous snakes and this will mean a general increase in such reptiles." Unfortunately, he lamented, "most people who go camping, hiking and otherwise vacationing out-of-doors are woefully ignorant of snake habits and habitats." In other interviews he urged Americans to wear shoes, long pants, and most importantly "puttees" [leather leggings] because they were "90 percent effective."[20] He believed that fewer Americans were bitten than elsewhere in the world, however, due to the fact that rattlers outnumbered all other venomous snakes in the United States and they warned before they attacked.

This scientific education about snakes occurred in no small part because of the expansion of zoos' snake collections during the 1920s and 1930s. Newspaper reporters covered the work of reptile curators in great detail, often with illustrative photographs depicting curators holding snakes or "milking" their venom. The coverage was clearly motivated by the fact that snakes, at least venomous ones, were potentially quite dangerous. Reptile curators played key roles in reducing this danger by helping to develop an antivenin distribution system housed, largely, in New Deal reptile houses.

Developing a North American Antivenin Program

Compared to people in other parts of the world, North Americans have had relatively little reason to fear snakes. Of the 115 species of snakes in America, only 15 (mainly *Crotalids*—rattlesnakes) are venomous, and even those are less deadly than some species on other continents. However, until the 1930s, America's medical profession had no widely available treatment for venomous snakebites, meaning that such bites annually killed scores of Americans and numerous pets and stock animals.

The incidence of venomous snakebites in the United States never rivaled India's, where in 1891 at least 21, 389 people — or one out every 10,424 — died from snakebites, but it was nonetheless sobering to many early twentieth-century observers. Those who studied snakebites calculated that mortality rates for snakebite victims at the turn of the century ranged from 10 percent of those bitten in the North to 35 percent in the Southwest, and they argued that the total number of annual bites and deaths was dramatically undercounted due to poor record keeping. Researchers had difficulty gathering data on more than a few hundred bites per year in the first few decades of the twentieth century, although by mid-century they could place the annual total at around 3,000 incidents, a figure still in line with recent Centers for Disease Control data.[21]

In the absence of scientific knowledge about how to treat snakebites, Americans, like people around the world, lived in reasonable fear of venomous snakes, a fear exacerbated by media coverage that emphasized not only the horrific physical consequences of a bite, but also the unpredictability of when one might become a victim. Newspaper readers would be forgiven if they received the impression that snakes were everywhere and no one was safe. During the 1880s and 1890s the *New York Times*, for example, featured a regular column titled "Bitten by a Snake," and ran many additional snakebite stories from all over the nation. In 1890, the *New York Times* reported on Pennsylvania's "epidemic" of blacksnakes, copperheads, and rattlesnakes that were "creeping into people's houses on farms and in the country towns, crawling into kitchens, coiling themselves up comfortably in cupboards and closets." Mrs. Oscar Boyer was made critically ill by a copperhead that had crawled into a barrel of potatoes in her cellar. Elsewhere it was not safe to go outside, as a "respectable" Montgomery, Alabama, couple discovered on a Sunday afternoon in September 1886 when their three children, ages two to six years old, failed to return from their play area near a fallen tree. The horrified parents found the "terribly swollen" children, all three victims of a deadly rattlesnake that had bitten them multiple times. On August 2, 1889, young Nellie Klugman was killed by a "huge rattlesnake" in her backyard near Volga City, Iowa. Professional and amateur berry pickers risked bites from snakes that hid under the bushes waiting to ambush mice that lived on the berries, as Frank Fritz learned while picking blackberries on a Pennsylvania mountain known as "Devil's Hole."[22]

The terror of the bite was rivaled only by the horror of the venom's effects. Before they died, victims of venomous snakebites experienced terrible symptoms, which the newspapers documented. After the initial pain of the bite, the parts of the body close to the bite started to swell and hurt. Next, victims developed a fever and their swollen skin started to change colors; within two or three days the skin would take on a blackish tinge as the flesh started to rot. If they had received enough venom, victims could be dead within a few hours. Prior to death, people could become unmanageable. William Reed, struck by a copperhead while bathing in a stream in Flat Creek, Tennessee, tried to bite people who attempted to treat his fever.[23]

Not surprisingly, given their ability to cause humans great harm, snakes were also killed. Snakes who slithered into homes might be met with force from protective housekeepers such as Mrs. Daniel Bush of Windgap, Pennsylvania, who discovered a "huge rattlesnake" in her doorway and "blew its head off with a shotgun." Sarah Sanford, who lived near Ellenville, New York, encountered a rattlesnake and a pilot snake in her hen house and cut off their heads with an ax. Semi-professional snake hunters killed snakes and offered them for sale pickled or as skins. An Ethiopian immigrant living in Connecticut caught a large number and variety of snakes that he then pickled and sold to bars in New York, Philadelphia, and Boston. Rattlesnake "roundups" and other less organized slaughters of snakes were relatively common. Texas waged a war on rattlesnakes in the latter 1920s in which bounty hunters could earn 35 cents a bag.[24]

One small group of Americans, however, wanted to protect snakes — even though it was at the greatest risk of snakebites: amateur and professional herpetologists, many of whom worked in zoos. These men and women frequently had close encounters with venomous snakes, including exotic species, impressing upon them the importance of easy access to antivenin. Although familiar with snakes and quite cautious, herpetologists were never completely safe, as evidenced by those bitten while collecting rattlesnakes or while taking care of snakes in zoos. In 1921, Ditmars received a cobra from Malay who needed surgery to remove his shedding skin from its body and eyes. The smooth crate used to transport the snake to the Bronx zoo had failed to give the snake the kind of rough edge needed to slough off its skin. Unfortunately for the snake, its new skin was coming in underneath the old and constricting its body. The old scales over the eyes were also blinding it. The snake's condition required that Ditmars somehow get into its cage to remove the skin and then capture it so that he could perform surgery on its eyes. The rescue mission was successful for both the snake, whom Ditmars helped to regain his sight and remove his dead skin, and for Ditmars, who was not bitten. A newspaper report of the danger that Ditmars faced, however, shows how much herpetologists at zoos needed a reliable cure for snakebites, If he were bitten, Ditmars' assistant would have cut into Ditmars' arm multiple times even before he administered one of the few vials of antivenin in the United States.[25]

William Mann faced a similar dilemma when the king cobra given to him by Ditmars for exhibition also developed scales over its eyes. Before the operation Mann consulted at length with Ditmars about how best to proceed. When news reached the public about Mann's impending operation on the snake, he received hundreds of postcards either applauding him or urging him not to operate because he ought not risk his life "for the sake of a snake." Luckily for the cobra, he decided in favor of the operation, arguing that "as one of his charges" the cobra "deserves whatever medical attention is indicated even at the risk of personal life."[26]

Mann, however, was also interested in the development of antivenin to treat venomous snakebites. On January 15, 1926, he and his fellow members of the Vivarium Society listened to a lecture by Martin Crimmins, a colonel in the army and amateur herpetologist who, as we shall see, was involved in the struggle to develop a domestic supply of antivenin in the United States. A society meeting the next year graphically demonstrated the necessity of this cause. The group met at the zoo to discuss the Brazilian production of antivenin, "the first shipment of which had just been received in Washington." There were "more than twenty members" at the meeting when one unnamed participant brought out a European sand viper in a crate to share. Although he handled the cage carefully, the snake popped out and slithered among the naturalists. Understanding the danger, Professor Brady, the head of the herpetology and biology department at George Washington University, caught the snake. Despite his use of forceps wadded with paper that was supposed to keep the snake from moving, the paper fell out of one side, allowing the snake to bite him on the left index finger. Brady began to give himself an injection of the antivenin (available to a small number of Americans at this point), but it required a fair amount of disrobing, which created "consternation among the women members of the society."[27] For propriety's sake, he left the meeting and went to a doctor who injected the serum. He was subsequently very ill with a fever of 103 degrees, but he recovered in a few days and returned to teaching. It was a nice coincidence that the society members happened to have the newly created antivenin on hand, and it impressed upon Mann the need to have widely available antivenin.

Ditmars also saw the need for antivenin for a range of exotic snakes. The necessity of a cure for poisonous bites was clear to him after he purchased a South American bushmaster for his own collection. He knew the behavioral patterns of some American rattlesnakes, but he was unprepared for the strength and striking strategy of this highly venomous snake. When it arrived, Ditmars placed the bag containing the snake into its intended cage. The snake's slight weight gave Ditmars the false impression that it was easy to handle, and he cautiously used a long stick to push the snake out of the bag. In a flash, the bushmaster turned around and headed for the cage door. Ditmars stepped back, but the snake was out of its cage and moving toward him with apparent aggression, backing him halfway around the room before he successfully used his staff to push it back into its cage. Soaked in sweat and much more aware of the danger of these snakes, Ditmars closed and locked the door.[28]

Marlin Perkins also suffered a near-fatal bite by a Gaboon viper (a particularly venomous African species). His assistant immediately sucked the venom out of Perkins' finger, but Perkins spent several weeks in the hospital recovering. One of Southern Bell's advertising campaigns in the early 1930s highlighted the speed of their long distance calls through Marlin Perkins' snakebite. After all, time was of the essence after a snakebite, so what could be more useful than a long distance call? The title of the advertisement that appeared, not coincidentally in three Texas papers — the *San Antonio Light*, the *Abilene Daily*, and the *Amarillo Globe*— was "This Man's Life was saved by a Long Distance Call *that went through fast*." The text read, "Emergency!

A voice shouted. Get Ray Ditmars quick...'" The advertisement told readers that "Molly Malone, a long distance operator, did not know the speaker was George Vierheller, Director of the St. Louis Zoo. She did know that nine minutes earlier Marlin Perkins, curator of reptiles had been bitten by a deadly Gaboon Viper." And Ms. Malone "did not know that Ditmars, curator of reptiles at the New York Zoo, is one of the few authorities on the treatment of American snake bites."[29] Speed was of essence in the treatment of venomous snakebites, and modern technology such as the phone, automobile, and airplane were key elements in quickly conveying information and medicine after a bite. Most important, though, for Perkins and other snakebite victims, was the growing availability of antivenin.

As a herpetologist and zoo curator, Ditmars' expertise was in collecting and exhibiting snakes, but in the 1920s he and other herpetologists joined forces with Martin Crimmins, doctors, and a Brazilian antivenin researcher to collect snake venom and develop a system for producing and distributing antivenin in the United States. Prior to the 1930s, wholly unscientific and ineffective snakebite remedies were used in the United States by most Americans, despite the existence elsewhere in the world of proven antivenin treatments. Nonetheless, a number of individuals worked to overcome legal, political, and logistical barriers to ensure that as America entered the 1930s, it would have a robust system of antivenin production and distribution. And they used zoos as the central storage place for antivenin because herpetologists at these institutions were frequently in danger and because they were better than medical doctors at identifying snakes — a crucial skill because there was not a single serum that treated all bites. Moreover, some harmless snakes that bit Americans looked like poisonous snakes, but really were not, in which case the person might only need treatment for a potential infection at the bite site. All of this required someone who had a very sure knowledge of snake species and that meant that doctors relied heavily on herpetologists at zoos. Knowledgeable doctors and herpetologists, however, had an uphill battle to fight against an American public that had embraced a range of home treatments for snakebites in often futile attempts to stop their devastating effects.

Into the early 1900s snakebites were treated with a range of ineffectual remedies. Many ordinary Americans resorted to folk cures that, while interesting, did nothing to ease the victim's pain or counteract the venom. Others relied upon painful treatments designed to stop the spread of venom. Some doctors, believing that research supported their treatments, sometimes actually accelerated the pace of the venom. This lack of medical knowledge about how to treat snakebites reflected America's isolation from the larger medical community, because researchers in other nations were busily developing effective antivenins during the first two decades of the twentieth century.

Folk cures reflected desperate attempts to save lives. When a rattlesnake bit five-year-old-George Putnam of Stony Fork, Pennsylvania, his parents cut a live chicken in two parts and placed the meat on the wound. They also found, killed, and cut open the snake, placing its dead body on the swollen bite marks. Later a physician cauterized the bite, but doubted that it would do any good. B. C. Coltrin, a miner working on the lower Rogue River in Oregon, was trying to catch his horse in a field when he was bitten by a large rattlesnake. He immediately bled the bite and bound tobacco around it. Hearing his cries for help, his neighbor Bud Fate took the additional precaution of covering the bite in coal oil and soda.[30] Coltrin lived, but it is unlikely that the tobacco, coal oil, and soda had much to do with his survival. Such measures posed no direct danger to bite victims, and since many victims received too little venom to kill them, folk remedies could easily gain anecdotal credibility.

A little bit of knowledge about how venom worked could prove dangerous. Recognizing that venom spread from the bite site through the rest of the body, some doctors sought to prevent the spread of the venom with cauterization or chemical agents. In theory, cauterization would work if one could instantly burn the wound and surrounding tissues because it would completely arrest the course of the poison. Not surprisingly, however, very few Americans were holding hot irons when they were bitten, and by the time a hot iron was ready, venom had traveled beyond the bite site. Such unfortunate patients experienced the pain of cauterization without any therapeutic value. Victims were also treated with a variety of chemical reagents including nitrate of silver, caustic soda, and gold chloride. All of these agents could annul the toxic power of venom outside the human body. Unfortunately, they were not successful when applied to actual bites. The most common chemical agent used to treat bites was permanganate of potash that sometimes helped the symptoms a little, but also destroyed healthy tissue in the process. It was also toxic to humans when injected at a rate that might have neutralized the venom.[31] In short, common methods for arresting the course of venom were likely as dangerous as the venom itself.[32]

A familiar treatment, endorsed by many American doctors, actually quickened the venom's effect. In the 1880s, a Dr. Charles Allen offered what he called scientific evidence that imbibing alcohol would both numb the pain of a snakebite and neutralize the venom's toxicity. Dr. Allen collected poisonous snakes from Florida, including the diamondback rattlesnake and the water moccasin, and brought them back to his home laboratory, where he conducted his experiments. He recommended whiskey for snakebite victims to "prevent heart failure," although he cautioned that most people would kill themselves if they drank a whole bottle. When William Gore, a 22-year-old stone cutter from Fort Lee, New Jersey, was bitten by a rattlesnake, House Surgeon Dunning at Manhattan Hospital gave him "an ounce and a half of whiskey an hour" so that he was "drunk all the time." The doctor also gave him carbonate ammonia to aid the whiskey in keeping the heart beating.[33]

A final set of measures had some basis in scientific research and continued to be recommended throughout the twentieth century, although not everyone agreed with the protocol. This procedure was designed to extract as much venom as possible and arrest its spread. First, a tourniquet was applied above the bite to contain the venom. Then several incisions were made at or above the bite site, and blood (which in theory contained the venom) was suctioned out by mouth or with a mechanical device. In cases involving animals, the flesh around the bite was sometimes actually excised. Experiments that involved drawing blood from bite victims (animal and human) and then injecting the blood into a healthy animal showed that the venom could remain potent for hours; thus, researchers firmly believed in the value of suctioning out as much blood as possible. Other experiments conducted on dogs in 1928 and 1929 indicated a much higher survival rate if these methods were employed. The methods did have some drawbacks, however. Tourniquets could be misapplied, resulting in gangrene and loss of limbs; incisions could become infected, especially if made in the field with non-sterilized blades; and suction could be applied too late or for too long, leading to significant blood loss. Also, because the fang is curved, the venom is injected some distance from the entry puncture site, making it difficult to determine where to cut and suck.[34]

The central weakness with all of the treatment methods employed before 1926 in America is that they did not directly attack the venom's toxicity. Yet, H. Sewall of the University of Michigan had conducted experiments in 1887 that showed how repeated small doses of rattlesnake venom could create immunity in laboratory animals. Less than a decade later on the other side of the world, at the Pasteur Institute's Bangkok station, Albert Calmette reported that he had created an effective antivenin.[35] Following Calmette's example, Brazil's Butantan Institute began collecting venomous snakes in 1910 to develop its own antivenin program. Pharmaceutical companies in the United States had less of a profit motive to manufacture antivenin, since there were relatively fewer venomous bites to treat. North Americans were thus behind in their treatment of snakebites, and this situation had been somewhat exacerbated by the federal government's lack of interest in the problem.

Historically, the United States government has played a large role in the attempted extermination of "undesirable" animals such as wolves, coyotes, birds of prey, and insects that threaten farmers or ranchers. Poisoning and trapping programs administered by the U.S. Department of Agriculture nearly wiped out many such animals. Venomous snakes, however, escaped the attention of government agencies, nonvenomous species were sometimes even protected by state law, and the occasional government study of snakebites underestimated the problem.

The American governmental agency that might have reduced snakebites at the end of the 1800s was the Bureau of Biological Survey, housed under the Department of Agriculture. The bureau exterminated many so-called pests to help farmers around the country. Their deadly war against birds, coyotes, and wolves is legendary. They did not include snakes, however, among the nuisance animals — in part because many snakes, including venomous ones, help farmers by hunting rodents and other small animals that eat their crops. Because "agricultural losses from the ravages of rodents in the United States closely approach[ed] a hundred million dollars every year," there were even laws on the books in states that prohibited "the killing of mammal eating, harmless serpents."[36]

What makes the lack of government interest somewhat puzzling, however, is that venomous snakes did appear to pose an economic threat to some ranchers. Poisonous snakes could take a fairly heavy toll on livestock, for example. One estimate is that in the mid 1920s, poisonous snakes accounted for over one million dollars in lost cattle in Texas alone. Horses were also common victims because they would lower their heads to investigate a rattling snake, resulting in a bite to the face where swelling could cause suffocation. When a rattlesnake bit Frank Corbett's horse on the nose, he gave his horse whiskey and brandy on the theory that it would keep the heart beating even as the poison attacked the rest of the body. Sometimes the victim pulled through in spite of the remedy, as in the case of this lucky horse whose head returned to its normal size after a few days.[37]

Despite the threat snakes posed to humans and domestic animals, it seems that the Bureau undertook only one formal survey of venomous snakes, and that study may have undercounted the number of fatal snake bites in the South and Southwest. Dr. P. Willson wrote a Bureau report on venomous snakebites of humans in the 1920s, but according to critics he failed to accurately estimate the number of bites in those regions.[38] He neglected, for example, to collect any bite data from entire states, including Mississippi. Contemporary critics argued that the report's numbers were so far off that they gave the false impression that snakebites were not a serious public health problem.

From the Department of Agriculture's perspective, venomous snakes represented a relatively minor threat to farmers, ranchers, or the general public. This lack of concern meant that snakes escaped wholesale government extermination efforts, but those who worked closely with snakes believed that action did need to be taken to protect the public from snakebites. Working through informal networks, these men facilitated the dissemination of information about snakebites, set up snake venom collection sites, and coordinated the importation and later domestic production of antivenin. The group included professional men from the military, the medical profession, and, especially important, zoos.

One military man in particular worked closely with Raymond Ditmars in supporting the creation of a domestic antivenin program. Colonel Martin Laylor Crimmins was the son of John D. Crimmins, a New York philanthropist. Martin was studying medicine at the University of Virginia when the Spanish American War broke out in 1898, and he volunteered and was accepted to Theodore Roosevelt's Rough Riders. After returning to the United States, he was eventually given command of Camp Bullis, Texas, in the 1920s. Colonel Crimmins was as also an amateur naturalist, so while stationed there he shipped poisonous reptiles to the Philadelphia, New York, and San Diego zoos as well as to natural history museums. He took a personal interest in the snakebite problem because he found that he had one rattlesnake bite for every thousand soldiers. He pursued bite victim stories in newspapers and came to the conclusion that there were likely more bites and deaths than commonly thought. Crimmins found that in 1928 alone, 130 Texans were bitten by venomous snakes. The available medical care at the time was poor, as he learned firsthand in 1925 when he was packing black-tailed rattlesnakes to ship to Ditmars at the Bronx Zoo and was bitten on the thumb. He tried to get help at the William Beaumont General Hospital in El Paso, but found that they did not understand how to treat poisonous snakebites. For example, when his pulse seemed to fade away, the doctors gave him adrenaline to restart it. Crimmins had worked with herpetologists like Ditmars and knew that adrenaline only accelerated the venom and that patients could recover on their own from a very low pulse rate. In fact he knew of patients whose pulses had been so low that they were undetectable, and yet they recovered within half a day.[39]

Because the death toll in Texas alone was so much higher than the small number of fatalities that the biological survey claimed for the region, Crimmins became convinced that he needed to get involved in the prevention and treatment of snakebites. While in Texas, he published "information about snake-bite and its proper treatment, and issued bulletins to soldiers and lectured in these schools, so that non-scientific people could distinguish poisonous and harmless species."[40]

In the mid–1920s, Crimmins, Ditmars, and other Americans turned to a Brazilian researcher for help. Brazilians had largely conquered their venomous snake problems by then, and at the center of Brazil's success story was Afrânio do Amaral, a protégé of Vital Brazil. While Americans were still using folk cures to treat venomous snakebites, Brazilians had translated Albert Calmette's pioneering scientific research into effective antivenin treatment. For his part, Calmette, a student of Pasteur, had created his initial serotherapy laboratory in French IndoChina after a nearby bombing war between France and China drove large numbers of cobras into the village of Bac-Lieu near Saigon. Forty villagers were bitten and several of them died within a few hours. The tragedy prompted Calmette to search for a snakebite cure, which he found by slowly injecting the venom of cobras into horse blood to allow the horse to build up immunity to the poison. From the horse's blood, Calmette then separated the antibodies, from which the antivenin was derived. Vital Brazil, also a Brazilian, read about and repeated Calmette's studies in the early 1900s at Butantan, a public scientific research institute outside of Sao Paulo. When Brazil left the institute, his protégé Afrânio do Amaral briefly became director before leaving in 1925 to take a position at Harvard University as an adjunct faculty member in the Department of Tropical Medicine.[41] Although do Amaral soon returned to reassume leadership of the Butantan Institute, he traveled frequently to the United States and worked closely with Ditmars, Crimmins, and others.

In 1927, do Amaral was awarded the Scott Medal and $1,000 by the Academy of Natural Sciences in Philadelphia for his antivenin research. Taking stock of his contributions, he distinguished his work from that of the British, who tried to reduce cobra bites in India during the same period by paying for cobra heads. The British system failed to dent the snakebite problem, and even by the early 1920s, the colonial government estimated that there were roughly 25,000 fatal bites per year. In contrast, do Amaral's protocol reduced snakebite deaths in Brazil from around 1,500 per year to only three to five cases by 1923. In do Amaral's opinion, the English destruction of snakes was as futile as "filling up a large container without a bottom."[42] By 1933, thanks in part to do Amaral's work, Americans had reliable access to antivenin manufactured by the Mulford Company.

The North Americans valued do Amaral because of his demonstrated success in organizing a multi-faceted program for reducing snakebite deaths in Brazil. First, he successfully convinced workers to adopt the simple prophylactic method of wearing high, leather calf leggings to protect their legs from bites. Second, he replenished the venom he needed in part by exchanging free vials of antivenin for either venomous snakes or snake venom that someone had milked. In this way he received free venom from a widening variety of venomous snakes that he would have been hard pressed to collect himself. Third, like a small number of other researchers around the world studying snakebites, he realized that one could not always cure all bites with a polyvalent serum. Initially antivenin researchers hoped that cobra serum would cure other snakebites. Instead, sometimes snakebite victims needed serum made from the venom on that particular kind of snake. And fourth, do Amaral perfected two techniques for producing antivenin — one fast and one slow. In the slow technique he extracted the venom and dissolved it in equal parts of a normal saline solution and glycerin. Next he gave horses subcutaneous injections of small doses of venom every few days for six months. By the end of that time a typical horse could resist a dose about 10,000 times larger than the initial dose of .05 milligrams.[43] In the faster process he injected venom neutralized by the pre-collected antitoxin from the slower process. Because the venom was already neutralized, he could inject larger quantities sooner to accelerate the serum producing process.

In a 1929 cover story, *Time Magazine* heralded do Amaral as the "soft-voiced suave herpetologist" and claimed that his work developing serum in the United States was "relatively easy."[44] The reality was more complex. Do Amaral enjoyed his travels to the United States, where he found a kind of scientific oasis and an intense interest in snakes. It is difficult to imagine the gains that Americans made against snakebites being possible without his help. He could not have done it on his own, however. During his visits to the U.S., he was helped by Raymond Ditmars, Martin Crimmins, and Thomas Barbour at Harvard University. Together with some additional colleagues, these central actors created an antivenin production and distribution network in the United States.

Impressed by do Amaral's and Vital Brazil's work, Thomas Barbour (1884–1946), who directed the Museum of Comparative Zoology at Harvard University, requested that do Amaral study the snakebite problem in the United States. Barbour — described as "good natured," "wonderful," and "quite profane" — was generous in both his proportions and his relationships with young scientists, standing over six feet tall and providing grants and publishing opportunities to promising herpetologists. Like so many university researchers, he enjoyed good friendships with zoo professionals, visiting the Philadelphia, New York, and National zoos each winter as he made his way to the warmer weather of Florida and the Caribbean. William Mann counted him among the best friends that he had made at Harvard.[45] In Barbour, do Amaral found a colleague who would open doors for him all across America.

After completing a trip through the American Southwest in 1925, encouraged by Barbour and "reviewing all available statistical data," do Amaral concluded that there were "more than 1,000 cases of [human] snake-bite by poisonous species" per year, rather than under 200 as the Biological Survey had mistakenly calculated. In a later publication he claimed that the number of poisonous snakebites in the United States was "more than 3,000 annually."

He found that the highest mortality rate from the bites occurred in the Southwest, where 35 percent of the victims died. Most of the Southwestern cases came from Texas and New Mexico. In Texas alone more than 150 people were bitten from July 1926 to June 1927. In the United States, do Amaral created North American antivenin sub-stations that in their first four months of operation in 1926 collected venom from over "4,000 rattlesnakes, as well as many from copperheads and moccasins." In addition, the presence of the antivenin substations led to greater reporting of bites. For example, of the 150 bites that he recorded in Texas, 50 percent of these occurred in the San Antonio district where he founded the antivenin substation with its efficient laboratory. Based upon the increase in the number of bite cases reported because of the antivenin station, he predicted that San Antonio's bite rate would soon exceed that of the state of Sao Paulo. The bites occurred in every part of Texas—arid land, forests, in the center of cities, and, as Americans knew, inside houses. Though other Americans have had an image of citizens in Texas perpetually clad in cowboy boots, the reality was different—do Amaral found that a significant reason for the high number of dangerous bites was that "the population of the state seems not to be used to wearing shoes or leg protection"; unfortunately "the majority of bites occur in the lower extremities." Young men were the most at risk. Data collected from the antivenin substations showed that 50 percent of the victims were 20 years old or younger, and 69 percent were male.[46]

Unlike Brazil, the United States lacked a functioning laboratory that could produce and sell antivenin, so under do Amaral and Barbour's persuasion, the Mulford Company (later Merck Pharmaceuticals) of Glenolden, Pennsylvania, agreed to create the antivenin serum for distribution in the United States. The company established the Antivenin Institute of America, and by 1929 it had antivenin available. The institute was necessary in part because Brazil was prevented from selling the antivenin produced at Butantan in the United States prior to 1925 since the U.S. Department of Public Health Service "prohibited the importation of serum from Brazil because it did not comply with its requirement of a test in living [human] bodies instead of a test tube." In addition, American public health authorities had "shut the door to [antivenin] importation into the United States because their rules [required] that any medical serum must be made under the supervision of United States inspectors—and there is no Federal Inspector on the ground to watch and approve of the Brazilian Government Laboratories." Do Amaral generously arranged for antivenin to be donated to the Bronx Zoo and other places in return for venom collected from zoo snakes, but such a solution was obviously short-term. As Ditmars pointed out, there was "an abundant supply of venomous snakes" that were just a few minutes from the main streets of the major cities in New York, Pennsylvania, and New Jersey," so the antivenin was sorely needed in his part of the country and elsewhere as well.[47]

The difficulty of securing antivenin in the United States meant that access to treatment was limited. In an interview with a St. Louis newspaper, Ditmars expressed pleasure that "we are already equipped with this anti-snake bite serum to some extent. We have it here in the New York Zoo." However, he stressed that "supplies are not yet sufficient or widely enough distributed to fully protect all the people." He promised that "within two or three weeks this situation will be remedied" because "we are bringing up from South America considerable quantities," to distribute "to such strategic points as will make it possible to reach and treat" snakebites around the country.[48] This was optimistic because each snakebite victim used two to three ampules of antivenin, but sometimes do Amaral would only bring "ten to twelve ampules" with him during a visit.[49]

Nevertheless, he promised the citizens of St. Louis that "within a year we shall have general and efficient distribution under the auspices of Harvard University department of Tropical Medicine, the New York Zoological Society, the United Fruit Company with its many steamers plying between North and South America and the Mulford Laboratories at Philadelphia which have been selected by Harvard University as distributors and will sell the serum, but not at a profit." Afranio do Amaral would "be in charge of the whole matter in a general way."[50]

Initially, Thomas Barbour established New York, Philadelphia, New Orleans, San Antonio, and San Diego as the antivenin distribution points, with substations to be named thereafter. At the time, proponents of the antivenin stations claimed that these cities were chosen by Barbour because they covered "the nation's snake-bite danger zones as indicated by the reports of all State Boards of Health." The substations were designed partly for geographical diversity, but primarily for their proximity to concerned members of the antivenin network, many of whom were at zoos. New Orleans, for example, was chosen because of the Audubon Zoo's collection of snakes and their zoo director's desire to participate. This was also the case in St. Louis, New York, and San Antonio, where either herpetology curators or zoo directors had previously traded venom to do Amaral in exchange for vials of antivenin. For example, when do Amaral visited the Bronx Zoo, he and Ditmars milked venomous snakes so that do Amaral could give the venom to the Mulford Company. Ditmars recalled handling "as many as a hundred during a couple of hours."[51] He used the same snakes repeatedly, however, because their glands would refill within a matter of days. Because each snake produced only a few drops of venom, he used many animals to reach an inch of poison in a drinking glass. All of the liquid was dried and stored for laboratory purposes at the zoo. It was easily reconstituted by dissolving the dried flakes in water and kept most of its original strength. After the venom was dried, it was poured into flat dishes and placed under a bell to avoid contamination. When the venom dried, it crystallized and flaked off somewhat. Working with the dried product was dangerous because once inhaled even small particles could produce some of the same symptoms as an actual bite.

Similarly, the Mississippi Valley's substation was the snake house at the St. Louis Zoo. Do Amaral taught zoo employees how to milk the poison when he visited.[52] As in New York, the reptile division in the St. Louis Zoo exchanged milked venom from their snakes for antivenin vials and lectures from do Amaral when he visited. The zoo kept the antivenin on their premises, but had it ready for delivery to local hospitals when needed. For example, the St. Louis Zoo had antivenin ready when a young boy was bitten by a poisonous snake and taken to St. Mary's Hospital.

The exchange sometimes worked in the other direction when zoo curators went to Butantan. Ditmars visited do Amaral in Brazil, and together they decided that they needed to know whether the venom from snakes freshly caught in the wild differed from captive animals. Although they knew that snakes in captivity had venom that had made perfectly good antivenin, they wanted to know whether the chemical make-up of venom changed at all with captive snakes. So, Ditmars returned with the goal of collecting rattlesnakes in the Berkshires and sending them on to Brazil.[53]

Meanwhile, Martin Crimmins devised a plan to fly patients to antivenin substations or take the serum to the patient using military planes. In California, Navy planes took serum to bite victims, and in Texas army aircraft "covered 2,600 miles hurrying the antivenin to the injured."[54] Poor people in rural areas were the most in need of rapid transportation to the nearest hospitals. On the whole, their emergency flight system worked, although there were times when they failed to rescue victims because they were too remote. A venomous snake bit an aspiring priest while he was vacationing at a summer camp at Port La Vaca, Texas. Although Crimmins tried to send the antivenin by airplane, he found that there was no landing field nearby and the roads were so poor that it would have taken eight hours to travel 150 miles by car. In this case the young man died.

Luckier patients, however, were picked up at airplane landing strips and flown to hospitals with antivenin supplies; or, zoological workers got them the antivenin. When a rattlesnake bit a man in Ithaca, New York, Ditmars' daughter sped vials of antivenin to Grand Central Station where the director of the New York Central *Nightflier* picked them up and shaved precious minutes off his trip up north to deliver the serum. From there it was taken by a driver who charred the car's floorboards speeding to the hospital.

Crimmins, Ditmars, and other Americans involved in the antivenin network also utilized the media to educate the public about how to avoid snakebites. Lacking help from the government, they conducted their own public health campaign to prevent bites. In an interview with Edward Marshall, a newspaper reporter, Ditmars warned Americans to "be very afraid of [snakes] this summer." He reminded the public that "many men and boys who were "not afraid of snakes" were killed from snakebites. He revealed that they had just rushed anti-snakebite serum to a "Boy Scout's camp several miles away to a lad who had not been afraid of snakes." They saved his life, Ditmars conceded, but "we could not save him from the incredible suffering snakebite causes." He reminded readers that "had it not been for the serum and the reasonably prompt administration, that boy would be dead today."[55]

Ditmars was particularly concerned about Boy Scout groups that showed little respect for the dangers of snakes. Boy Scouts during the 1930s collected what they called mini natural history "museums" during their camping trips that they exhibited at the camps. Snakes were among the trophies that the boys caught for this purpose. Ditmars warned the boys that although the snakes seemed docile, they could strike in a moment. He noted that "familiarity breeds contempt" and that snakes handled safely "nineteen times might bite the twentieth time." The "only safe [venomous] snake for a camp museum," he concluded, "was a dead snake."[56]

Ditmars had specific advice about how to prevent bites. In addition to fearing snakes he recommended avoiding stone walls where rodents lived. He also warned that "low shoes and sheer stockings for girls in snake-infested neighborhoods" were "tempting providence."[57] Not even boys, he noted, should wear low shoes when in tall grass or the forest. Finally, he warned that antivenin supplies were insufficient to protect all victims.

Although the antivenin stations saved some patients, they struggled to keep enough serum on hand. At one point, Ditmars nearly ran out of antivenin after giving much of his stash to bite victims in the Southwest. Similarly, although do Amaral gave Crimmins over $2,000 worth of antivenin for the stations, Crimmins began running out in the latter 1920s because of a rash of snakebites. Seeking to bolster his supply, Crimmins made the unusual decision to experiment with making himself into a producer of antivenin by injecting himself with repeated small doses of the poison to become "*in person* a stock for emergency cases." Laboratory experiments indicated that his blood could be injected into bitten animals to save their lives, but Crimmins gave so much blood to these studies that "he used to come into his own office pale and spent." Doctors in San Antonio, Texas, actually used his immunized blood twice on humans. In the first case a child came into the hospital, and the doctors gave her three hours to live. "The Colonel offered his blood for transfusion; the child lived seventy-two hours."[58] In the second instance another child was bitten and given Crimmins' blood. This time the child lived. On the same day that the second child survived, the United States Public Health Service authorized the production of serum in the United States on the grounds that it had been tested in a human. Crimmins' experiment came to an end.

Even as antivenin was starting to become available in the United States, Americans continued to use ineffective treatments. Despite the hard work of zoo herpetologists and do Amaral, antivenin was considered just one of the possible treatment methods. Do Amaral noted in frustration, for example, that Americans continued to get drunk after a snakebite even when they knew that antivenin was available at the substations. They got so drunk that they were "not able to go to the hospital or a clinic to get the antivenin injection." He believed that cauterization was helpful if "done under the proper circumstances," but unfortunately it was "often as bad as the poison because of the lack of septic application." Or they held on to the belief that potassium permanganate could cure snakebites. A common over-the-counter snakebite kit contained only that ingredient. More commonly, prior to receiving antivenin injections, most patients had their bites incised and suctioned, either in the field or in a hospital. According to do Amaral, however, "sucking the blood at the wound is useless" because the virulent elements in the poison integrate quickly with the tissues where the fangs have struck and the suction of the lips cannot pull it away." And some experts who knew a great deal about the power of antivenin serum continued to advocate other completely useless measures as well. Crimmins, for example, recommended colonic irrigations for bites.[59]

Even before snakebite victims reached an antivenin substation (or any hospital), they were sometimes treated with a Dudley bite kit. This kit was named after Dr. Dudley Jackson who, together with Dr. R. Rhea (a veterinarian) and Crimmins, conducted about 700 live experiments on animals partly to debunk myths about the curative ability of home remedies, and partly to demonstrate their theory that cuts over the wounds and suctioning could remove poison and save lives. Jackson and Rhea conducted extensive studies on dogs that they injected with venom to compare the survival rates for no treatment, incision, and incision and suction. The lucky dogs survived; the unlucky ones died. The research convinced Jackson, Rhea, and Crimmins that incision and suction were critical life-saving tools. Crimmins went so far as to insist that snakebite victims should undergo an hour of suction before the administration of antivenin.

Jackson, Rhea and Crimmins recommended making cuts over the bites to withdraw the venom through a suction device. If that failed to work, they

SNAKE-BITE

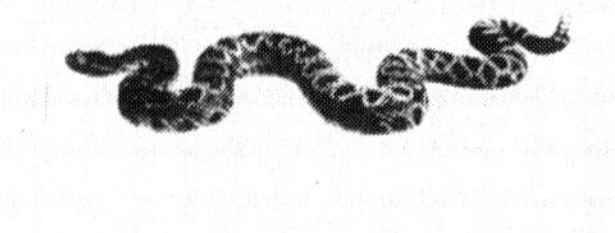

RESTING COIL

CRAWLING POSITION

STRIKING COIL

HOW A RATTLESNAKE BITES

FANGS FOLDED

ADVANCED FOR ACTION

STRIKING

MODEL HEAD, WITH SKIN REMOVED. SHOWING RIGHT POISON GLAND AND DUCT LEADING TO ONE OF THE TWO HYPODERMIC NEEDLE-LIKE FANGS.

SNAKE-BITE TREATMENT

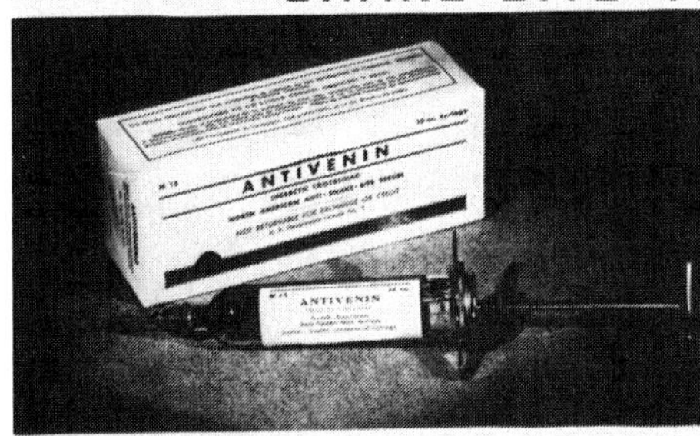

SNAKE-BITE SUCTION OUTFIT

1. TOURNIQUET 3. RAZOR BLADE 5. LARGE APPLICATOR
2. ANTISEPTIC 4. SMALL APPLICATOR AND SUCTION BULB

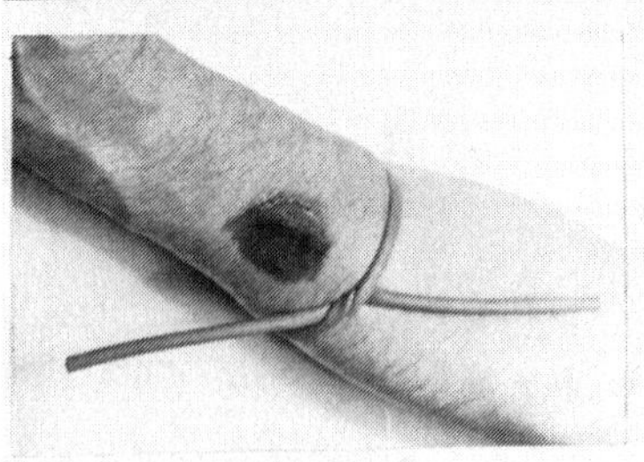

PREPARATION OF WOUND FOR TREATMENT

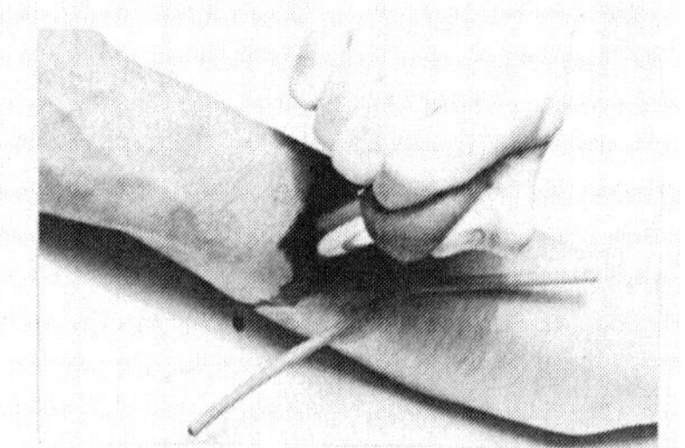

REMOVING POISON WITH SUCTION BULB PRIOR TO ANTIVENIN INJECTION

PHOTOS BY L.M. KLAUBER; AMERICAN MUSEUM OF NATURAL HISTORY. SHARP AND DOHME. RALPH DE SOLA. FEDERAL WRITERS' PROJECT

Mulford antivenin snake-bite kit. Zoo herpetologists played key roles in collecting venom, raising public awareness about snake-bite treatment, and distributing antivenin (NYC Municipal Archives).

proposed making even more cuts further up from the injured spot on the body. For example, if one was unable to withdraw venom from a cut over a bite on the lower part of the arm, then they advocated making several cuts in an *x* shape on the upper portion of the arm near the shoulder and suctioning the venom. In 1929 Jackson claimed that San Antonio hospitals had treated 17 cases of poisonous snakebite by suction alone with no deaths.[60] And finally, they advocated tying a ligature above the wound to slow the pace of the venom.

As do Amaral had warned, there were serious problems with all three parts of the protocol. They neglected to emphasize, for example, that suctioning venom worked best in the laboratory when it was done within about five minutes of a bite. Similarly, making small incisions with a sterilized knife was relatively safe. Unfortunately, most Americans typically did not carry around sterilized surgical equipment. The Dudley snakebite kit included a sterilized blade and suction device, but many Americans lacked access to the product and resorted to using whatever sharp object they could find, only to survive the snakebite but die from tetanus infection.[61] And finally, ligature only makes sense if one understands how to use it, and if one can get to a hospital quickly. Unfortunately, poor people living in remote, rural regions of the country were tying ligatures around their arms and legs only to have gangrene set in because it took them so long to get professional medical treatment.

Some of the potential problems created by the treatment advice advocated by Crimmins and Jackson are evident in the bite Ross Allen received. Allen was a professional snake collector and venom dealer who owned the Florida Reptile Institute. He was on a routine collecting trip when he lost his footing while trying to hold a five-foot diamondback rattlesnake. The snake succeeded in biting him on the thumb. After he caught the snake, he applied first aid by placing a tourniquet on his arm and by cutting his hand open in three places. Luckily for Allen, friends had accompanied him to get the snakes and thus drove him toward help. They sped through the woods to avoid the deep sand they had gotten stuck in while traveling to get the snakes and then caught the road for the remaining 27 miles to Ocala, Florida. The tourniquet failed to stop the venom which rapidly spread through his arm and into his body. In 30 minutes he was paralyzed and unable to use his hands or legs, so his friends made three more cuts, applied suction and injected 30 cc of antivenin. When he reached the hospital, he was "totally helpless and only able to talk, in spite of the first aid" he received. At the hospital the doctors made 50 incisions from his hand to his shoulder, but his arm continued to swell and the pain remained intense. A special nurse "kept up suction" for nearly "24 hours." He had lost "considerable blood" from his first aid and

Snake house built by WPA for the San Diego Zoo (1937) (New Deal Network, www.newdeal.feri.org).

needed transfusions.[62] Although he received morphine, it failed to entirely dull the pain. Within 24 hours the swelling spread from his arm to his shoulders, his back, chest, and head. By the fourth day, doctors had to remove part of his thumb because gangrene had set in. Muscle spasms in his legs, face, and stomach afflicted him for six days. Ultimately Allen survived, but the part of the treatment with the fewest negative side effects was clearly the antivenin.

In the zoo world, 20-year-old Roger Conant learned a similar painful lesson on September 16, 1929. New to the Toledo Zoo and eager to build its snake collection, Conant had dreams of persuading the Mulford Biological Laboratories to establish a venom-collecting station at his zoo. Recognizing that he would need to learn how to extract venom from rattlesnakes, he consulted with the St. Louis Zoo's reptile curator, Marlin Perkins, who advised that he always milk with his left hand, thereby keeping his right safe from the ill effects of a bite. Dutifully following this wise suggestion, Conant nonetheless was bitten on his left thumb by a speckled rattlesnake (*Crotalus mitchelli*). Coming to Conant's assistance, a doctor with a nearby office applied the tourniquet, cutting, and sucking procedures. Unfortunately, the cut he made on Conant's thumb was so deep that it nearly severed it. At the hospital, Conant was given a shot of the Mulford antivenin, but no tetanus shot. The upshot of a horrific few weeks in the hospital was the amputation of his left thumb.[63] The case illustrated that the risks from medical mistakes could be greater than the danger of the venom itself.

The role of American zoos and, to a lesser extent, the U.S. military, in antivenin production and distribution was significant. As we have seen, reptile curators at major zoos collected venom and provided it to antivenin researchers and producers. As antivenins became available, they were typically purchased or acquired by zoos, which kept them on site and provided them to local and regional hospitals as necessary. Of 31 antivenin sources listed in a 1980 book about snakebites, only six were not zoos. More important, at that time, the only comprehensive survey of antivenin sources had been compiled by the American Association of Zoological Parks and Aquariums, and a permanent "punch-card file of available antivenins" was kept at the New York Zoological Park (the Bronx Zoo). Thus, Dr. Henry Parrish, a national expert on snakebites, recommended that in emergency situations hospitals should contact local zoos or the Bronx Zoo to locate antivenin, which could then be airlifted by the U.S. military to any location.[64]

Speaking to Army Medical Corps Officers at Fort Sam Houston in 1927, one of the few non-zoo antivenin stations in the United States, do Amaral "pointed out the need for State or county aid in establishing antivenin stations throughout Texas" because the mortality rate from bites could reach "35 percent of persons struck." He was imagining stand-alone laboratories devoted entirely to preventing death by snakebite. Although his idea made sense, the reality of American social policy is that it tends to graft improvements on pre-existing programs and institutions. Widows Pensions preceded Aid to Dependent Children, the central poverty program under the New Deal. Civil War pensions were a kind of model for Social Security.[65] In the United States herpetologists at zoos had initially offered their institutions as the center of venom collection and the holding place for the substance once collected. Like other social programs in the 1930s, the idea for antivenin substations came from a previous decade, but New Deal funding to zoos indirectly expanded the program. It is striking to note that without the work of herpetologists at zoos, members of the American military, and a government-paid Brazilian scientist, thousands more Americans would have died if the private medical sector had been left to develop and distribute antivenin on its own.

By the end of the 1930s, zoological professionals had made serious attempts to raise the status of snakes and other reptiles and to educate the American public about their virtues. They used New Deal money to create prominent buildings that satisfied the substantial interest that Americans had in reptiles. Research into antivenin had made Americans safer. If this had been a decade of zoo construction, it had also been golden age for reptile curators. Yet, the snake itself was still at risk.

As America prepared for entry into World War II, some herpetologists began to cast snakes' virtues in patriotic terms. Carl Kauffeld, for example, asked Staten Island Zoological Society members to remember the positive features of rattlesnakes. They had brilliant, lidless eyes that suggested vigilance, but they never began an attack; however, once engaged, they never surrendered. "As if anxious to prevent all pretensions of quarreling with the weapons with which nature favored her, she conceals them in the roof of her mouth, so that to those who are unacquainted with her, she appears most defenseless." Rattlesnakes also generously alerted their enemy by sounding their rattles. In this way Kauffeld hoped to shed light on why early Americans selected the rattlesnake on their flag and why they should think about her positive characteristics for the looming war. Quoting Rheua Vaughn Medden's *Tales of the Rattlesnake: From the Works of Early Travelers in America*, he wrote that "the power of fascination attributed to her by a generous construction may be understood to mean that those who consider the liberty and blessings which America affords, and once come over to her, never afterwards leave her, but spend their lives with her." The rattlesnake resembled America in that she is "'beautiful in her youth and her beauty increases with age; her tongue is also blue, and forked as lightening, and her abode is among impenetrable rocks.'"[66]

An editorialist at the *New York Times*, however, was still skeptical about the educational ambitions of zoological herpetologists. In 1942 five hundred Staten Island children attended a reptile demonstration at the Staten Island Zoo, prompting the editor to worry about the potentially false impression that the herpetologists gave them. "As an adventure for the children, and perhaps as publicity for the Staten Island zoo, the incident may have some merit," the writer opined. But when it came to the "educational or social value of the experience a person has his doubts." He insisted that it should "hardly be the purpose to teach children never to pass judgment on a snake until it has been determined whether the reptile is poisonous or not" because that was "carrying fair mindedness a little too far." Instead, he believed that parents of small children preferred to have them "keep away from all snakes, even at the risk of hurting the feelings of the harmless kind." Besides, he editorialized, we needed to have a sense of good and bad that snakes provided. Snakes for the author were plainly bad and so if we failed to instill that fact in our children then we risked the very end of our democracy. Herpetologists were unintentionally confusing American children by showing the positive attributes of snakes so they were susceptible to the same kinds of arguments in favor of dictators. And, they might be convinced by critics of democracy, causing them to lose their "flaming faith." How could people fight for the United States, he wondered, if they "insist on being cool about everything and fair to everybody, even to the snake in the grass."[67]

For zoo professionals in the 1930s, being "fair" increasingly meant protecting animals, largely from ignorance. Through their extensive publications and availability to the media, reptile curators tried to build public sympathy for a maligned animal at the same time that they protected the public. But education alone was insufficient. Most zoo animals needed protection in a more immediate sense — from substandard structures that were detrimental to their welfare. The New Deal made it possible for zoos to exhibit reptiles in state-of-the art buildings that incorporated the latest knowledge about how to keep snakes alive and healthy. This interest in improving the quality of cages extended to other areas of America's zoos. The average visitor to a New Deal zoo may have been impressed by how the new structures created a more pleasurable viewing experience. Zoo directors too appreciated the recreational benefits of new construction, but for many the New Deal enabled another noble goal: the development of healthier, more humane exhibition techniques. For these directors, the New Deal was not just for people; it was also for animals.

FIVE

A New Deal for Animal Welfare

When the St. Louis Zoo opened its reptile house in 1926, unexpected controversy raged around the fact that the predator inhabitants ate other animals. Humane Society members and outraged citizens protested the "inhumanity" of allowing the snakes to prey on live rabbits and pigeons, sometimes in front of spectators. From the perspective of the reptile curators, such protests irrationally put the welfare of the rabbits and pigeons above the welfare of the animals on exhibit — the snakes. In fact, the curators intended to feed the snakes dead food in the future, but to ensure that the snakes developed good eating habits, they began them on a more natural diet of live prey. The curators knew from experience that pythons, for example, were attracted to the movement of prey and therefore were more likely to eat and remain alive if presented with live food. Later, curators shifted to dead rabbits, but shook them to give the illusion of life and movement to attract the snakes. Unfortunately, the brutal honesty of this feeding display was not appreciated by all spectators. The zoo, Humane Society, and local newspapers received "scores of protest" about the "reported feeding of live rabbits and birds to the snakes." The Humane Society secretary took about "2000 telephoned complaints" about the feeding. Some visitors were so angry that they told him that they "disapproved of having snakes in the Zoo at all." In her letter to the editor of the *St. Louis Star* protesting the reptile house, Mrs. Charles Dolan expressed outrage at what she believed were the misguided funding priorities of the city. She noted that the city had "complete indifference and neglect of the city's horses" as evidenced by its "failure to subsidize the Humane Society" in its work to rescue fallen horses. For Mrs. Dolan the welfare of horses took precedence over the welfare of captive snakes. In general agreement about the relative value of various animals, the Humane Society lodged a formal protest about the Snake House, leading the zoo to discontinue the live feedings and switch to a diet richer in rats than in rabbits and birds.[1]

Other zoo directors defended the feeding of live animals to snakes on the grounds that not only was it natural, but critics of the practice were hypocrites who lacked true compassion for animals. In a zoo newsletter article about "feeding the snakes," Staten Island Zoo Director Clyde Gordon reluctantly omitted an image of an indigo snake eating a live frog because "some well meaning person, while sitting at the dinner table enjoying roast lamb and all the fixings, might exclaim, 'Isn't that horrible — a snake eating a poor little frog?'" To further expose the hypocrisy of such people, Gordon related a story about his visit to the home of a well-educated, upper-class woman who had complained about live mice being fed to snakes. Gordon discovered her home alone, her husband off on a quail and pheasant hunt. While he was chatting with her, the woman's house cat entered with a mouse, which it tormented until dead. Finally taking his leave, Gordon mockingly admired the woman's fur coat, made from "100 or 150 gray squirrel skins."[2]

These episodes revealed the uncertainty about what, exactly, indicated a concern for animal welfare. From the reptile curators' perspective, feeding captive snakes live prey demonstrated the highest regard for the reptiles' welfare. To the Humane Society activist, the inhumanity of allowing an innocent rabbit to be killed and eaten outweighed the welfare benefit to the serpent. In a broader sense, awareness of the impact of the New Deal on the health and well-being of captive exotic animals existed just below the surface of public consciousness, rarely rising above the enthusiasm for the new zoos' tremendous entertainment and educational value. Zoo directors and even some humane societies, however, were keenly aware of the fact that animal welfare standards were a significant issue in the design and maintenance of New Deal zoos. Concern for animal welfare was not a universal value in the 1930s, but it motivated many of those with a professional interest in zoos. This chapter explores the place of animal welfare in the construction, staffing, and media coverage of zoos during the era.

In comparison to the average zoo of the previous decade, New Deal zoo projects represented major transformations in the standards of care for captive wild animals. Dilapidated structures with design flaws that actually threatened animals' health were replaced with "modern" buildings and enclosures that featured technical innovations to promote sanitation and design concepts intended to improve animals' mental health. Thus, permanent construction materials such as concrete, brick, stone, and tile, enclosures without bars, heating and ventilation systems, and naturalistic exhibit decorations became New Deal norms. In cities that stopped short of a full zoo remodeling, New Deal laborers undertook smaller projects intended to improve the animals' welfare. The *New York Times* concluded that nationally "federal funds have made [zoos] better for their inmates," citing zookeepers who "offer evidence that the New Deal is favorably affecting animal psychology." New York parks landscape architect Allyn Jennings offered a spike in lion cub births as "proof" that the New Deal cages in the Central and Prospect Park zoos had improved the animals' comfort.[3]

Zoo professionals did not rely only on exhibit design and maintenance to promote animal welfare. Of equal and perhaps greater importance was the quality of the zookeepers. Experienced and knowledgeable men with a fundamental sympathy for the animals in their care were highly valued. The ability to have frequent and close contact with animals was a sign of one's skill. Thus, circus men moved easily into the zoo world, both at the keeper and manager levels. Where 21st-century zoos eschew any circus connections, the 1930s zoo often collapsed the distinction between the two. And as we shall see, Humane Society workers sometimes worked closely with zoo directors.

Unfortunately for the welfare of the animals, neither New Deal funding nor the presence of a zoo director supported by a local Humane Society guaranteed that a city would significantly improve the condition of its zoo. The decision to commit funds to CWA and WPA zoo projects was ultimately a political one initiated at the local level. The case of Seattle's Woodland Park Zoo, which saw only modest gains during the New Deal, will be examined in this chapter to illustrate the complex political dance that could determine the fate of a city's animals.

Finally, the welfare gains resulting from the New Deal must be considered against an era where the popular media treated animal deaths and injuries not with concern or even indifference, but with actual amusement. Escapes, injuries, and deaths were the stuff of entertainment to reporters.

Animal Welfare

Animal welfare was an integral part of the zoological world in the 1930s. It differed, however, in degree and emphasis from the kind of welfare that later generations came to espouse. There was no sense that individual animals should have formal legal protections regarding their care and housing, which the American Animal Welfare Act (1970) eventually afforded. Nor did zoological directors believe that they could save endangered animals through breeding, as their later colleagues hoped. Nevertheless, zoo directors, keepers, and curators employed a rhetoric and took actions that suggested a commitment to humane principles of animal care. Foremost for directors was the conviction that cages should be as clean and safe as possible for the animals and their keepers. With New Deal funds at their disposal, directors dreamed of repairing or even replacing outdated and inadequate exhibits. Those with the physical space for expansion could even plan for modern, bar-less enclosures of the kind designed by Carl Hagenbeck. By emphasizing the inhumane conditions of their existing zoos, directors could justify repairs and new construction.[4]

Even the "best" zoos in the late 1920s had physical problems that could harm animals. Wooden animal houses were firetraps that held up poorly to the effects of gnawing, urine, and weather. Many structures built in the nineteenth century were not only dilapidated, but also poorly conceived. When Roger Conant took a position as curator of reptiles at the Philadelphia Zoo in 1935, he concluded that the old lion house had been designed by a "moron." Each cage had a slanting hardwood floor so that a keeper could hose the cats' excrement and urine to the back of the cage, sloshing it through a slot into the basement below where the keeper then shoveled the mess into buckets for removal. Newly hired director Marlin Perkins arrived at the Buffalo Zoo (New York) in 1938, walked into the main zoo building — an old wooden structure with double-decker animal cages — where his office was to be located, and promptly threw up in reaction to the stench. Chicago's Lincoln Park Zoo Director Alfred Park complained about poorly constructed exhibits that were "impossible to keep clean and free from disease." In Seattle, the zoo director in 1933 reported "the animal cages were in deplorable condition."[5] New York's Central Park Zoo was grossly antiquated. Park Commissioner Robert Moses described the zoo as a "disgrace," so unsafe that the keepers carried guns so "they could shoot the animals in case the [wooden] buildings caught fire." The buildings also lacked adequate lighting, heating, ventilation, drainage, and water, making them "unhealthy and unsanitary, as well as much too inadequate for housing wild animals."[6] One observer remarked that "the whole place smelled so strong [of urine] that the wise went there only in the winter." Moreover, Tammany corruption had created a patronage culture where employees' private desires were put above the public's and the animals' needs. One restroom attendant had reportedly removed all of the plumbing so that she had space to install a grand piano.[7]

Conditions were so bad according to some external critics, that there were calls to close zoos. For example, between 1924 and 1934, the New York City Parks Department had made some efforts to improve the Central Park Zoo, but they had been sporadic and uncoordinated, the onset of the Great Depression curtailed significant park work, and political support for the zoo was lukewarm at best. During the 1920s, the *New York Times* editorialized several times that the dilapidated menagerie blighted Central Park and should be removed altogether. In response to such threats, the Department of Parks *Annual Report* for 1929 described several projects: an enlargement of the seal tank, the construction of a "winter stone hut," minor repairs and painting of building interiors and exteriors, and the installation of electric blowers "for the purpose of disinfecting and ventilating the Lion House." But by 1930, the zoo's annual budget was only $20,280, with no funds spent on physical improvements. A 1931 *Regional Plan of New York and Its Environs* recommended removing the zoo to Wards Island and demolishing the adjacent Arsenal Building, in which the Parks Department was headquartered.[8]

Similar complaints were repeated by zoo directors and concerned citizens in city after city. Nowhere, it seemed, was there a modern, humane public zoo in the late 1920s. Given the fact that many of these zoos were less than a decade old, the cynic might suspect that tears shed on behalf of animals were disingenuous, that zoo directors cared less about the welfare of their animals than about the prestige and power that could come with a larger, more modern zoo. Several facts weigh against such an interpretation. First, directors had been expressing concern about the condition and design of their zoos well before New Deal funds became available. In part, zoos suffered because of budgets that provided for salaries and food, but not for regular maintenance or new construction. Buildings, especially those constructed largely from wood, simply wore out quickly. In addition, they were not always designed with the animals' welfare or the keeper's convenience and safety in mind.

Zoo design principles established in the late nineteenth century and still followed by many zoos in the first two decades of the twentieth century tended to satisfy human aesthetics more than animal welfare. Building exteriors were made to resemble classical temples and civic structures, Swiss chalets, Turkish temples, or Renaissance pavilions — all designs that tended to limit natural light. Expensive ornamentation delighted visitors but added nothing to the structures' utility. Mistaken beliefs about proper care also influenced design. Tropical animals were housed indoors permanently, shut off from the supposedly dangerous fresh air of northern latitudes. Ventilation was nonexistent. Animals did not always have any area to which they could retreat from view.[9] Paints contained lead. Wood floors absorbed urine. Visitors could generally view animals quite easily, but keeping the cages clean and preventing the spread of disease could be a challenge for curators. For smaller municipalities, cages could be quite minimal, as little as some fencing and a modest shelter or even a cage set on bare earth. The luxury of an animal "building" that housed multiple species was beyond the smallest zoos. In short, directors who bemoaned "inadequate" and "disgraceful" structures had legitimate complaints, and local humane societies often agreed.

Innovations in technology and design, many originating in Europe, highlighted the weaknesses of old-fashioned zoo structures. With the opening of the Hamburg Zoo in 1907, the profession was exposed to Carl Hagenbeck's philosophy of the "barless" zoo, where hidden moats rather than fences separated viewers from the animals, who roamed in relatively expansive open spaces that featured artificial rocks and other naturalistic elements. Anti-German sentiment resulting from World War I, however, delayed the adoption of Hagenbeck's principles in America. By the 1920s, reinforced concrete became a preferred material, favored because of its suitability for fashioning into naturalistic shapes and its ease of cleaning. Wood floors, for example, were replaced with concrete ones that could be hosed down. Plaster walls were tiled. Zoo architects also began designing exhibit spaces that took advantage of natural light and ventilation. Vita-glass skylights let through more ultra-violet rays than ordinary glass.[10] In sum, architects and zoo professionals began trying to solve welfare problems through better materials and design.

The Audubon Zoo's monkey island, constructed by the WPA. A common feature in New Deal zoos, these islands were popular with visitors and were advocated by zoo directors, who touted them as improvements in animal welfare. The playground equipment gave the monkeys a way to exercise and amuse visitors at the same time (courtsey of New Orleans Public Library).

The typical New Deal zoo incorporated some or many of these welfare-oriented innovations, leading to hyperbolic press releases claiming the construction of the "most modern" zoo in the nation. Such claims were clearly exaggerations, but they did get at a fundamental truth. The new zoos' exhibits usually represented significant improvements over the old ones. Wood construction was replaced with brick, stone, and concrete. Even a tiny zoo such as Indiana's Washington Park Zoo featured stone buildings, giving them at least the impression of permanence. Local materials salvaged from old buildings or even quarried on site enabled cities to provide materials at little or no cost, with the federal government picking up the tab for the labor. The Toledo Zoo dramatically expanded in this way.[11] The liberal use of concrete led to a profusion of "monkey islands," unfenced island mountains on which various monkey species caroused, restrained only by a water moat. Concrete was also used to create exhibits that were modern in both zoological and architectural senses. Following the lead of their European counterparts, architects in Buffalo and San Francisco designed, respectively, a seal pool and a monkey island that eschewed naturalistic elements in favor of abstract forms. Resembling nothing from the animals' natural environment, these forms at least had the virtue of being easy to clean. A few zoos with space to spare — the Detroit and Cincinnati and Ft. Worth Zoos — built Hagenbeck-style exhibits.

Directors happily emphasized the welfare benefits of new exhibits. The Cincinnati Zoo advertised its Hagenbeck lion grotto with a photograph of a pleased lion and the invitation, "Come up sometime and see me in my barless cage. I'm home most all of the time. Enjoy a wholesome day in a healthful atmosphere."[12] Here the animal himself attested to the healthy condition of the New Deal exhibit. This was no cage, but a "home."

New Deal laborers made other improvements that affected animal welfare. Vermin infestations were exterminated. Drainage systems were improved to keep paddocks free of standing water. Hazards such as exposed electrical wiring were removed. Fences were made dog-proof. Infirmaries, research labs, commissaries, and administrative offices were built. These kinds of projects were barely visible to visitors and received almost no media coverage, but they made it easier to care for zoo animals. They represented investments in zoos' basic infrastructure.

Given the rhetoric about animal welfare that typified so many justifications for New Deal zoo projects, it might come as a surprise to modern observers to realize just how many of the "modern" exhibits were in fact quite simple cages, with concrete floors and walls, and a mesh or barred face. For example, New York's Central Park and Prospect Park zoo buildings looked fantastic as buildings, but the actual animal cages were shockingly small by today's standards and not always well designed. Mischka, a mischievous

Central Park Indian Sun bear, escaped several times from her supposedly "escape-proof" cage.[13] For most animal welfare activists today, the more natural and expansive a zoo exhibit is, the better. "Cages," by contrast, with their iron bars, steel mesh, and concrete floors too closely resemble prison cells to be favored by either activists or zoos today.

But for zoo directors and curators in the 1930s, there was such a thing as a "good" cage. A clean, light cage housing a healthy animal that received careful attention from a sympathetic keeper was evidence that there was nothing inherently inhumane about cages. Commenting on San Diego Zoo's gorilla cage, director Belle Benchely argued, "That cage is their security; they obviously feel at home there. They have had food, shade, water, bedding, security, companionship, variety of interests, room for play and exercise, and protection from predatory beasts—all that they sought and received in the jungle." When an architect who designed a building for monkeys and lions asked Raymond Ditmars' opinion of it, Ditmars responded that "the heating pipes in the monkey section were too low" because too much heat so close made the primates lethargic. "The cages needed vertical mesh panels at the margins to prevent savage animals injuring unsuspecting specimens adjoining them. The skylights were too small. There was no independent ventilation of the cages. There wasn't enough slope to the floors to drain them. The doors of the cat cages, leading outside should always slide sideways, instead of dropping as they might hit an animal's back."[14] For Belle and Ditmars, both humane zoo professionals working at the most modern zoos in the nation, cages *per se* were not problematic. Poorly designed and inadequately maintained cages, however, presented real dangers to their inhabitants and keepers, and a key legacy of the New Deal zoo projects was minimizing those dangers.

Zoo Personnel, Animal Protection and Enrichment Within the Zoo

Low acreage, unimaginative park departments, or small budgets could all restrict the size and scope of New Deal zoo projects, so directors often settled for relatively small cages that protected the animals from fire and other hazards. Zoo directors accepted these smaller enclosures, however, because animal welfare for them was more about how humans interacted with the animals than it was about the size of their cages. It did not matter so much that a monkey had a small pen, as long as there was a human there every morning, afternoon, and evening to take care of the monkey by talking to it, feeding it, and frequently touching it. From their perspective, the quality of a captive animal's life was determined as much by its keeper's skill and sympathy as by the design of its enclosure. This kind of animal welfare philosophy was premised upon a profound faith in the potential kindness of humans and the belief that humans and animals could successfully interact with one another. Thus, in addition to ensuring safe and sanitary cages, zoo directors and curators believed that animal welfare meant caring for animals in as tender a manner as possible with as much human interaction as possible. From their perspective, animal welfare for many of their charges meant the time they spent outside of the cage either walking around the grounds on a leash held by a keeper, giving rides to small children, or hand-raised in a manner similar to human children. Panthers, and other large cats, were walked daily by keepers, zebras and hippopotami gave rides, and baby animals were raised by zookeepers' wives. Because the direct care of animals was the purview of keepers, they were the people who generally mattered the most in many ways. Consequently, there is an emphasis in the 1930s on the good people who cared for the animals.

By what standards was the quality of a zoo worker measured? A very few, such as Mann, could lay claim to academic qualifications, but a far more common sign of a capacity to work in the zoo was one's personal connection with animals. Even Mann proved this point. His textbook knowledge, while impressive, paled compared to his experience in the field capturing and caring for wild animals. A zoo man (or woman) was different from other men not just in his technical ability to care for exotic creatures, but also in his love and concern for those animals. Running a zoo was not a job. It was an avocation. But where could a boy or young man learn to care for elephants, tigers, snakes, exotic hoofed herbivores, and the many other foreign creatures that a good zoo stocked? The answer was simple. Go to where the animals were — the circus.

There were certainly more circuses than zoos in early twentieth-century America, and they had exotic animals literally by the truckload. Training and caring for animals was essentially a 24-hour-a-day job, and former circus men developed skills that translated easily into the zoo world. For example, Alfred Parker joined the circus at age 14, imported animals from Africa and Asia, and ran his own trained leopard show until, at age 44, he joined the staff of Chicago's Lincoln Park Zoo, where he served as director from 1919 to 1931. The National Zoo's head keeper, William Blackburne, had been a lion tamer with the Barnum and Bailey circus before he joined the zoo in 1890; he worked there until 1944, earning the highest respect of his more scientifically-trained boss, William Mann.[15] Gus Knudson ran away with the circus at an even younger age — ten — and spent his youth on the road, learning the skills that enabled him to hold the directorship of Seattle's Woodland Park Zoo from 1922 to 1947. Louis Vehellier, an accomplished zoo-man, ran the St. Louis

Zoo. The Cincinnati Zoo's longtime director began his animal career in the circus. These men, and the many others who worked as keepers, were valuable because they knew how to care for wild animals outside of cages.

Raymond Ditmars referred to these folks as "animal clairvoyants." Although many of the best animal workers originally worked in circuses, Ditmars believed that such people could come from a range of backgrounds. In his experience, foundry workers, painters, and iron workers had all demonstrated a special gift with animals. They were "at home with any kind of animal, savage or gentle, large or small." The key to their success was that they "quickly struck up an acquaintance with the animals, and had a strong influence on their charges." The young Marlin Perkins felt sure that a zoo career was right for him when the capuchin monkeys assigned to his care "became my friends." Cincinnati Zoo Director Sol Stephan advised young keepers to speak in kind and soothing tones with the animals to gain not just their trust, but also their love.[16]

A keeper with the right sympathy for animals was ultimately more important than a modern exhibit space. For example, Ditmars describes visiting an unnamed zoo in the western United States where the monkeys lived in a converted barn. One entire side of the barn was covered with greenhouse glass, which gave the space a tremendous amount of light that the animals enjoyed. On the other hand, the space was difficult to heat, although Ditmars theorized that the cooler temperatures were actually better for this species, as evidenced by the fact the zoo had bred more than any other more sophisticated zoo in the East. Ditmars believed that a "quarter of a million dollars could be spent for a monkey palace," but this more antiquated structure staffed by the "extremely sympathetic fellow" was better for the animals. Without any irony, Marlin Perkins could describe the "close personal friendship" between a keeper and an orangutan named "Spitting Sam." Jack Frazer, a Prospect Park Zoo elephant keeper, paid a visit to the new Central Park Zoo home of Chang, an Indian elephant he had once nursed through a sickness. "Handlin' animals is handlin' animals," he said, "but, Chang, now, me and her are friends."[17]

Those keepers without such sympathy could be downright dangerous. The Central and Prospect Park zoos were very similar to many other city zoos of the era in the sense that they were managed by a Park Department employee with limited expertise and understanding about wild animals. The zoos shared a director, Captain Ronald Cheyne-Stout, who perceived zoos' central public purpose to be recreational entertainment and took every opportunity to play the clown. As noted earlier, Stout enjoyed entering the cages while occupied sometimes at night to demonstrate his fearlessness — he claimed that cast-off circus lions were "sissies" — and the papers reported gleefully on the lost fingers and assorted maulings that occasionally resulted. On one occasion, seeking to impress the famous animal trader Frank Buck, the visiting Berlin Zoo director, and a crowd of 200, he picked up his blacksnake whip and entered the lions' cage after dusk and against the keeper's advice. Moments later, he hastily retreated but not before his backside was severely clawed by his pet "Curley." Whenever one animal killed another, which happened with some regularity in the small and crowed cages, Stout laughed off the incidents, remarking after one that it was "unavoidable." Stout also engaged in ill-considered stunts such as housing two mountain lion cubs with a domestic dog, an experiment that led to a protest from the Humane Society. Robert Moses called him an "irresponsible fellow" whose unprofessional behavior, "neglect of duty and failure to keep department records properly" finally led to his ouster and replacement with a better-qualified man, Dr. Harry F. Nimphius, a veterinarian. The choice of a veterinarian clearly signaled the zoos' commitment to welfare, since few zoos at the time had a veterinarian on staff, and many relied on the services of a local doctor or Humane Society veterinarian.[18] A veterinarian's ability to care humanely for animals was beyond question.

Zoo directors and curators also prized the qualities of cleanliness and scientific knowledge about diet as additional indicators of humane treatment. Joseph Ellis, a prominent animal dealer during the 1930s, and a founding member of the American Association of Zoological Parks and Aquariums, had a strong record of successfully bringing back animals alive. Ditmars attributed this in part to his cleanliness. Ditmars describes watching Ellis clean the animal dishes. He said he "rubbed them inside and out with his fingers, rubbed into all angulations and corners, rinsed them, and those receptacles were *clean* when filled and placed in the cages." In addition to good hygiene, proper food was crucial, even though keepers often had only trial and error to guide them. At an unnamed zoo in the South, for example, one keeper's knowledge about what to feed animals made a strong impression on the young Ditmars. This keeper decided some newly arrived and scrawny deer needed handfuls of rich "clover ... selected from the open bales, and rolled oats without chaff, instead of the ordinary ground oats fed the larger-hoofed animals."[19] They also needed "a bit of powdered charcoal mixed in" that he believed helped their intestines. At other times keepers hand fed the animals, pushing their food to them "bit by bit, with the tip of one's finger" to fend off starvation. St. Louis zookeeper Cash Ferguson brought the zoo's infant orangutan home each night to sleep between himself and his wife.[20] Photographs of keepers' or directors' wives bottle-feeding infant animals as though they were human babies were ubiquitous by the 1940s.

It could take a long time to gain an animal's trust, so patience and persistence were other key traits of the best keepers. San Diego Zoo Director

Benchley described getting to know Katjeung, an orangutan from Surabaya. His cage came with a sign that warned people away from the danger within, but Benchley slowly got to know the animal. At first she spoke to him quietly, "calling his name" and edging closer to his cage. After awhile he calmed down and looked at her "more intently as though listening to and studying my voice and expression." She held out half of a peach pressed to the wire for him. He took the fruit and retired to study it. Benchley made subsequent visits to his cage and gave him additional offerings so that "day by day" she gained his trust.[21] She knew she had won his trust when one day she came by and he followed her across his cage. She continued to visit him until he showed that he watched for her and enjoyed her visits. The happy and important consequence of this attention from Benchley and other keepers was that Katjeung's appetite increased. Thus, his health was determined not by the architecture of his cage, but by the quality of his connection with humans.

When Martin and Osa Johnson — a famous husband-and-wife explorer and nature film team — visited Benchley to present two young gorillas that her zoo had purchased, Martin informed her that he did not care for zoos. He further claimed that he never wanted to bring animals into captivity, although this view fit awkwardly with the fact that the Johnsons kept caged animals in Africa, where they used them to create scenes for their movies. They also sold animals to other American zoos. So, how do we explain Martin's comment? His view no doubt owed something to his and Osa's close relationship with wild animals. They kept them as pets and even traveled with them, so it seems improbable that they objected to zoos simply because they kept animals captive. Rather, captivity without appropriate care was the problem, and for the Johnsons and others the quality of care depended on the degree of trust and contact between the keeper and animal. After visiting the San Diego Zoo and seeing its staff in action, Martin informed Benchley that he liked her zoo and found it "the most humane in the world."[22]

Interestingly, Benchley was herself somewhat ambivalent about circuses. On the one hand she saw chimpanzees and other animals as intelligent beings in themselves, and she resisted the practice of dressing them to look like humans. On the other hand, she believed that some zoo directors had their animals' welfare at heart when they dressed them and made them perform because it gave them a job to do during the day, provided interaction with their fellow animals, and afforded significant psychological stimulation. Recalling her visit to the St. Louis Zoo, run by George Vierheller, she insisted that her belief that animals should not be dressed as humans is "not prompted by any belief that training and acting makes them unhappy, for no one could possibly look upon the large group of six to nine trained chimpanzees at the St. Louis Zoo, while they are putting on their show under the excitement of the cheering crowd, while they are being trained, or when just sitting around on exhibition in their fine home, and fail to see how important they feel and what fun they have during the performances." The chimpanzees danced, rode, and performed musical numbers that culminated in a "wild tandem pony race." The animals were obviously competing against each other during the race and were participating with a great deal of enthusiasm. The chimpanzees, moreover, added to the act each day with "extemporaneous tricks." She concluded that it was "evident to anyone who understands their nature that the devotion between the men and animals is mutual."[23] From what she could see, animal performances were not inhumane. Quite the opposite, they gave the animals joy and facilitated positive relationships with humans.

Nevertheless, Benchley profoundly disliked the practice of dressing up chimpanzees for acts. She admitted that she did "not care as much for trained-animal acts" as she "should perhaps." She believed that the "things that animals do when left alone are so much more cunning and smart," but unfortunately, people insisted on dressing them. Clothing the animals made them "pitifully short of being noble, attractive and interesting." She believed that dressing animals to imitate humans simply underscored their lack of similarity to them and gave them a "hideous, pitiful human likeness" that was created by the outfits. She believed that the garments concealed "all of the real fineness and nobility found in most wild creatures."[24] Benchley expressed a view that gained little traction in the zoo world of the 1930s but foreshadowed what the future would bring.

No better example of a sympathetic New Deal zoo caretaker existed than former New York Governor Alfred E. Smith, "night superintendent" at the Central Park Zoo and "booking agent" at the Staten Island and Prospect Park zoos. To support a sympathetic view of the Central Park Zoo, the Park Department hired artist Lilian Swann to illustrate a colorful children's book—*The Picture Book Zoo*—for which Smith authored a foreword. Here Smith enthused that the cheerful pictures "introduce the wild beasts so pleasantly that instead of bogies they might almost become friends" for children, just as they were for him.[25]

The quality that allowed Smith to make friends with wild animals had been eloquently described by his protégé, Franklin Delano Roosevelt. At the 1928 Democratic National Convention, FDR nominated Smith to be the party's presidential candidate. In his speech, FDR described Smith's various political virtues and then focused on the "one thing more" that would make him "a great President" in the line of Lincoln and Wilson. This one thing more was "that quality of soul which makes a man loved by little children, by dumb animals ... the quality of sympathetic understanding of the human heart, of real interest in one's fellow men."[26] This "sympathetic understanding" that led the Central Park Zoo's animals to trust Smith and welcome his nightly visits was the same quality that zoo directors looked for in their keepers and curators.

The publicity surrounding the former governor's warm relationship with New York's zoo animals led one West Coast animal welfare activist to approvingly (but mistakenly) praise Smith's leadership as the Central Park Zoo's director. The mistake revealed the effect that even a sympathetic "night superintendent" could have on public perceptions of a zoo's welfare standards. By objective standards the Central Park and Prospect Park zoos were architecturally beautiful but not very spacious. The former was frequently called a "postage stamp" zoo, new and clean, but quite small. Even its proud father, Moses, admitted that the Prospect Park Zoo had bigger cages, and that was not saying much. But, in the 1930s world, the smaller zoos could reasonably say that size did not matter. In New York the combination of Smith and a zoo director with a veterinarian degree gave the strong impression that the animals, however cramped, were loved and well cared for. This basic story held true from coast to coast.

Working with the Humane Society: Seattle's Story

Despite the enormous popularity of zoos, a few critics questioned the morality of keeping animals captive for human pleasure, especially in conditions that appeared inhumane. In 1939 W. J. Davies wrote Seattle's *Post-Intelligencer* to complain not about the conditions of the Woodland Park Zoo, but about "the principle of the thing" itself. Impugning the intelligence and moral development of the average zoo visitor, he argued that "anyone who can view the sunburned rump of a baboon with the vacuity of the simple is just one jump ahead of the anthropoids." He urged citizens to consider the "bedraggled" eagle "huddled" in his cage as "the emblem of our glorious country—Alcatraz in perpetuity, no reprieve, no parole." The "imprisoned" wolf deserved the "peace and solace of the death chamber."[27]

More typically, however, when local humane societies and anti-cruelty leagues turned their attention on zoos, they did so as allies of a kind. On the whole, animal welfare activists were more concerned with animal abuse in other areas of American society such as the remaining ill treatment of horses used to pull carts or the vivisection of animals for scientific experimentation. In contrast, zoos could be helpful partners, taking in animals that ended up in shelters and buying condemned horses.

Seattle, Washington, was home to some relatively active animal welfare advocates, with whom the Woodland Park Zoological Gardens director, "Dr." Gustaf (Gus) Knudson, enjoyed a fairly congenial relationship. Each year, for example, the zoo loaned animals, provided carts, and gave talks for a children's pet show and parade; Knudson emphasized that fellow organizers included the president of the State Humane Society, the president of the King County Humane Society, and local Humane Society officers. Aside from the occasional defensive reaction to a complaint about the zoo, Knudson, in his own words, "endeavored to work in harmony with both the State and County Humane Societies." He claimed to welcome "constructive criticism," no doubt because he was skilled at using animal welfare activists against the Park Board to lobby for changes he desired in the zoo.[28]

Hired in 1908 as an animal keeper, Knudson claimed a degree in veterinarian medicine, although it appears more likely that he was self-taught and had taken only a relevant course or two in veterinary work and dentistry. Born in 1884, the youngest of nine brothers, Knudson ran away from his Alexander, Minnesota, home at age ten when the Jim Lemon Circus came to town. The freckled, red-haired boy never looked back, enjoying life on the road with animals and developing a reputation that led to a job with the more famous Barnum circus. During these formative years, the plain-talking Knudson learned "don't never be afraid of animals; be kind to 'em," lessons that guided his work in the zoo. When he retired from the zoo business in 1947, he confided to a reporter that his favorite animal of all time was one of the first that he cared for in Lemon's circus — an elephant named Ruthie. He became the zoo's official director in 1922.[29]

The most potent sign of Knudson's ability to provide humane care to his charges was the "Dr." in front of his name. On his business cards, Knudson identified himself as a "scientist," a zoo director, and "a friend of all birds and animals." His scientific credentials were on display to all visitors in the form of the "operating" room off his office where shelves of medicines and surgical instruments testified to his medical expertise. As one reporter remarked about this office, "everything was as complete for its purpose as the office of the average M. D. to whom troubled humanity takes its ills." Such professionalism was not evident in all zoos, where a tiger's ingrown claw and a camel's abscessed tooth could be removed without even a local anesthesia. But as a self-proclaimed veterinarian, Knudson could treat injured or sick animals humanely and skillfully, as when "poor little Lena" (a monkey) got a hollow tin pencil stuck on her toe, which had to be amputated. Knudson's compassion for animals' welfare was also evident in his observations about the best and worst of zoo life, and he insisted "most of the animals in captivity die from neglect."[30]

Knudson never missed an opportunity to educate the Park Board and the public about the quality of care that he provided for Seattle's animals. In his first annual report to the Park Board he acknowledged that the Woodland

Park Zoo collection "cannot compete" in size and variety with those in New York or Washington, D.C. but insisted that "it does compare well in the excellence of its exhibits and the care given them." He went on to give concrete examples of the improvements he had made to the welfare of the zoo's kangaroos, bears, big cats, and monkeys. And he described methodical efforts to prevent the spread of disease among the animals. Finally, in stating the "object of the Zoological Garden," he emphasized welfare as the first priority: "to exhibit animals under favorable conditions." In a typical monthly report from the 1930s, he described an AAZPA reporter who had made a "thorough inspection" of the zoo and "remarked about the healthy condition of our animals," a compliment that Knudson claimed was made by "numerous persons" from "other sections of the country."[31] In other words, his zoo compared favorably with the nation's other zoos.

Were caged wild animals treated "more brutally" than domestic animals? Knudson thought not, pointing to the contrast between a caged bear and a "domestic bull with a ring in its nose." In fact, in a zoo "the first consideration should be the care and life of the animals," and Knudson fought for this priority both by wrangling over civil service rules that prevented him from keeping "trained" staff on hand at all times, and by lobbying for better accommodations for the animals.[32]

During his 25 years as director, Knudson voiced significant reservations about his zoo to the press and his board. In one such incident, for example, a lioness killed one of her young. Her name was Erabell and a newspaper reporter claimed that she was so maddened by the crowds that came to see her baby and so wanted to prevent it from living a caged life, that she killed it. According to the reporter, she "slew it to spare it the life she herself [was] doomed to lead behind the iron bars of captivity." The reporter, however, came to that conclusion because Knudson offered it as the explanation. According to Knudson "it was because she loved her cub that she killed it" and she would "kill the other too, if she is disturbed" because she could not "bear to see her loved ones lead a captive life."[33] Knudson's explanation of the death served two purposes. On the one hand, it exonerated him from blame, since the death resulted from the lion's natural reaction to captivity, not from a keeper's error. Moreover, Knudson wanted a new space for the large cats, so he had an interest in portraying the present cage as too confining.

In late 1926, Knudson made a formal appearance before the Board of Park Commissioners to report that the zoo was in "deplorable" condition. After his appearance the board sent out a delegation to investigate the conditions and determine if additional money should be appropriated for the zoo. As public discussions about the zoo continued, the King County Humane Society described the "present accommodations [as] entirely inadequate, especially from a humanitarian point of view," and recommended that the smallish zoo be relocated to a larger site. Rather than just "patch up the old" buildings, the city needed to build an entirely new zoo. The Humane Society emphasized that the present fault rested with the antiquated structures themselves, not with the zoo staff. Similar suggestions in the past, however, had gone unheeded by the Park Board, which always pleaded a "lack of funds." If the city could not find the funds this time, the Humane Society was prepared to call for the disposal of all the zoo animals. Such threats prompted a new survey of Woodland Park to study the possibility of zoo relocation, but the board took no steps to address the fundamental complaints.[34]

By July of 1927, Knudson, "discouraged with trying to keep the animals healthy under present adverse conditions," had effectively joined forces with the Humane Society to try to force some action. In an investigative report requested by Knudson and released that month, the Humane Society described the animals "in very bad condition" and "living in a deplorable state." Knudson himself claimed that "most of the animals are on the verge of nervous breakdown," confined to "hot, crowded pens." The report formally recommended that the zoo be relocated to a larger site and that an independent Zoological Society be created to "take charge of the city's zoological gardens." The latter proposal clearly expressed both Knudson's and the Humane Society's long-standing frustration with the Park Board's inability or unwillingness to adequately support the zoo. An independent zoological society composed of "citizens interested in the proper care of the animals" would presumably do a better job of locating and committing the resources that could give Seattle "a good zoo, not a mere makeshift" one. The report hinted that a small admission fee would help raise funds.[35]

The Park Board president's response indicated that no substantial changes would be forthcoming. Explicitly rejecting the idea of moving the zoo, the president noted the zoo's $21,000 budget and said, "we have spent too much there already." No new building would be necessary, he concluded, because "the animals are cared for as well as in any other zoo."[36] In a literal sense, the president was probably correct — most of the nation's city zoos were outdated and dilapidated in 1927. Other cities, however, acknowledged these deficiencies and at least paid lip service to plans for improvement, although it took the infusion of federal funds to turn those plans into reality.

To be sure, Seattle's Park Board took some small steps to address the Humane Society's and Knudson's concerns. During the next few years, the aviary was repaired, the big cats' building received "first-class heating equipment," and a commissary (for the slaughter of feed animals and the storage and preparation of food) was built; but the board recognized that "there are still many improvements to be made at the zoo for the purpose of maintaining the animals in a comfortable condition." Acknowledging the ad hoc and fundamentally inadequate nature of the existing zoo, the board recommended to the mayor and city council that a study for the "development of a *permanent* and up-to-date zoo" should be made.[37]

Towards this end, in 1930, the Seattle Board of Park Commissioners paid Knudson to travel to St. Louis, Kansas City, Washington, D.C., Philadelphia, New York, Boston, Toledo, Chicago, and Toronto to study "zoo construction and to secure detailed information of the construction of the housing units used in these zoos, particularly bear grottos, feline houses, bird houses, and other features." They were particularly concerned that he "study the housing conditions from a sanitary, medical, and humane point of view." During his 31-day trip, Knudson inspected 26 zoos and a number of aquariums and private animal farms. His report emphasized that animal exhibits were planned carefully to provide proper lighting, ventilation, and heating. Moreover, "all of the zoos of any importance" gave much attention and resources to care of their animals through auxiliary buildings such as "well equipped Commissaries, Bakeries, Kitchens and in particular Hospitals and ... supplies." He pointed out that every zoo had a director and either employed a veterinarian or kept one on contract; perhaps in a bid to assure his job security, he provided data revealing that the combined low-end salaries for these positions exceeded his own by several thousand dollars. Animal welfare did not come cheaply, and Knudson emphasized that "the main financial support" for these zoos came "through taxation for the zoo as any other City Dept." The cost of bringing Seattle's zoo up to the standards of these others would be high. Knudson estimated the 25-acre zoo would require an expenditure of hundreds of thousands of dollars to make a truly "modern" facility in which the current "prisoners" could experience the relative freedom of bar-less enclosures and naturalistic habitats. Knudson's report included an architectural rendering of a proposed Feline House that would exhibit the big cats in barless, exterior dens but allow them to retire indoors to heated quarters during inclement weather. Park engineer E. R. Hoffman's ten-year plan of 1931 reiterated Knudson's wishes but estimated the cost of such improvements at two million dollars.[38] Given the zoo's frugal $33,000 annual budget, such a modern zoo looked out of reach.

As the Depression gripped Seattle, plans for a better zoo were put on hold, but Knudson's strategy highlights the paradox that some zoo directors faced in trying to get better accommodations for their animals. Such directors were forced to claim that their institutions were bad for animals in order to compel politicians to commit public resources to their improvement. Apparently, only shame or the threat of scandal motivated these officials to invest in the zoo. Such attitudes could make unlikely allies of animal welfare advocates and zoo directors. Thus, Knudson frequently turned a formal complaint about conditions at the zoo into an opportunity to both assure his Park board of his own commitment to animal welfare and lecture them about the need to further improve the zoo. In 1930, Dr. Ira Brown, president of the King County Humane Society, expressed concern to the Park Board about whether the zoo's elephant, "Wide Awake," suffered from her confinement indoors during the winter months. In his reply, Knudson pointed out that Wide Awake was "neither chained nor hobbled" while inside, giving her a "freedom" that he had not observed at other zoos. Nonetheless, he admitted that the elephant house was too dark, and described his proposal to both enlarge the building and also add skylights. In closing his report to the Park Board, he wrote, "I am glad to note that members of the Humane Society are observing our animals and conditions for their welfare, and I am pleased to co-operate with them in an effort to make surroundings better each year."[39] Knudson's cooperative message and tone seemed intended less to appease zoo critics than to prod the Park Board into action; in this case the Board agreed that the suggested improvements were warranted.

Surplus and Slaughter

In the meantime, Knudson had to cope with a more prosaic problem: how to feed the numerous other mouths at the zoo. Some zoo animals reproduced naturally, and others aged to the point where they made poor displays. Over time, a zoo could end up with many more animals than it had the room or desire to exhibit. Larger zoos such as San Diego's might have scores of species available for sale or trade at any time.[40] Such unwanted specimens were called "surplus," and in good economic times, there might be private citizens, circuses, or other zoos willing to purchase them. Keeping surplus stock on hand was expensive. Nearly half of the Woodland Park Zoo's annual budget went towards feeding, so the board sought various ways to reduce those costs. While the board gave directives limiting the number of animals the zoo could acquire and reducing the numbers it already held, Knudson pursued a different strategy, working with the Humane Society to ensure a steady supply of low-cost meat for his animals.

One logical cost efficient measure was to stop the flow of animals to the zoo. Zoos typically exhibited multiple specimens of the same species, and they also received large numbers of common American animals from well-meaning citizens. Each of Knudson's monthly reports detailed multiple such donations. A few elite zoos such as the Bronx zoo educated Americans about what kinds of animals belonged in their bounds and turned away would-be donors.[41] However, not all could be so selective as the Bronx Zoo. In other places around the country, zoo directors often took in even common domestic animals. Such collection building was fine when the economy was strong but posed a problem when it was not.

In its cost-cutting quest, the Park Board came up against the King County Humane Society, which occasionally relied on Knudson's expertise and facilities to care for animals. The society brought a Shetland pony to the zoo "for treatment and care." Apparently, however, they assumed that the zoo would care for the animal for free. The Park Board disagreed and resolved that "animals are not to be accepted by the Zoo for treatment and care unless arrangements are made for remuneration for this service."[42] In addition, they made it clear that the society would have to get their approval before attempting to bring animals in for care.

Similarly, the Park Board grew weary of owners leaving pets at the zoo and later retrieving them only to claim that the zoo had failed to adequately care for them. For example, Mrs. G. F. Coy appeared at a Park Commission meeting "with an injured monkey, which she had presented to the Zoo about a year ago." She claimed that the monkey "was not given proper care by the animal keeper." The commissioners responded by passing a motion that thereafter, "animals, birds, or other property" would be accepted by "the Zoo only with the approval of the Board, and when so acquired" were the "sole property of the Park Department, and are not to be loaned out to the former owner or to any other person for any purpose without the consent of the Board." A few months later the board reiterated its ban on Knudson taking any more "birds, animals or reptiles" without first securing their approval.[43]

A more extreme and substantial efficiency involved disposing of surplus animals, thereby reducing the number of mouths to feed. The shortage of buyers, however, affected this option, as the Central Park Zoo discovered in 1930 when its annual surplus auction yielded $600 for a camel but failed to even attract bids for its four lions; two years later the entire auction netted only $28.60, in comparison to an $1,800 average over past years. Despite the adverse market, in December of 1932, Mr. Williams, chairman of the Zoo Committee on the Seattle Park Commission, "suggested that the common and unnecessary animals and birds at the Zoo be eliminated in the interest of economy." President Lustig considered the proposal and agreed that it was "advisable to dispose of certain uninteresting and superfluous specimens." At their February meeting, the board directed Knudson to eliminate "some of the unnecessary animals at the Zoo, either when there is an over-supply of certain specimens or when the animals are too common and uninteresting to be an attraction."[44]

Knudson complied with the board's wishes, as he reported rather graphically: "As ordered I have begun disposing of some of our surplus stock—pigeons, goats, a buffalo, and bears have been fed to other inmates." This may not have been exactly what the board had in mind. When they said that they required that animals be "disposed," they seemed to imagine that the animals would be loaned out to individuals or groups and then brought back as needed. As they said, "Dispose of such animals, as it may be advisable to recover later, when the stringency of the present time has passed, so that either the original stock or an equal number of younger stock may be returned to the Zoo when so requested by the Park board." If the zoo could not get its original animals back, they at least hoped to get their offspring. A few citizens did in fact offer to provide shelter for the animals. The zoo received letters from B.D. Shaffer and S.F. Fix, for example, who both offered to take the animals and then return them or their offspring when the fiscal crisis had passed. Similarly, Bill Lancaster sent a letter to the board requesting a parrot, which he promised to return when the zoo could afford to take care of it. This practice of placing animals in foster care was apparently successful enough that the board decided to take bids on the animals so that they could get as much money as possible for them. For example, they took bids on "domestic fowl or water fowl" from the zoo.[45]

Knudson preferred to keep his animals, so in the face of the board's calls for reductions in numbers, he emphasized the substantial cost savings in food that he generated by picking up surplus fruits and vegetables from Seattle's grocers and, more important, running an onsite slaughterhouse to provide low-cost meat to the zoo's carnivores. In this latter venture, Knudson was greatly assisted by the local humane society. Like other urban humane societies, Seattle's gave much of its attention to horses, still used during the 1930s to pull wagons and for farm work. Disposing of old and injured horses was one of the organization's central tasks, and the Woodland Park Zoo's slaughterhouse was a key partner, since the King County Humane Society had a formal contract with the Park Board to supply the zoo with condemned horses and cows at $10 and $7.50 a head respectively. The zoo's onsite slaughterhouse produced a sizeable amount of low-cost meat for the carnivorous animals. In 1933, for example, the Woodland Park Zoo slaughtered 121 animals, which produced 42,647 pounds of dressed meat. Some of the slaughter stock — largely injured or old horses — was donated by private owners, some of the stock (including bear, buffalo, and elk) came from the zoo itself, and some was purchased at a penny per pound. Horse-drawn wagons and even cattle shared urban spaces with automobiles, street cars, and pedestrians. The inevitable accidents were opportunities for the zoo with a slaughter house. When a bull escaped from the White Center Dairy and ran amok in West Seattle until felled by 20 police bullets, the dressed meat "made a good feast" for the zoo's big cats. A terrible accident at the Olympic Riding Academy resulted in three race horses ending up on the zoo menu.[46]

The slaughterhouse reveals the very different role of animals in society during the 1930s, as well as the dissimilar attitude of the era's animal welfare organizations. Humane society members in the 1930s were clearly not enemies of zoos or even circuses. The director of the Everett, Washington state, Humane Society and city pound, Happy Day, nicely illustrates this point. Until the Depression hit, Day had run the Happy Day Shows. In 1929, he traded several surplus animals to the zoo, including "Happy," an ironically named Giant Rhesus monkey with a "bad disposition." Knudson had to keep Happy in a separate cage, so he was quite pleased when Day returned in 1933 and offered two slaughter horses in exchange for the monkey. As Knudson noted humorously, "everyone concerned was happy" with this trade. Happy Day later went on to work for the Portland, Oregon, Humane Society.[47] The relationship was presented as unremarkable, but 75 years later it would be almost unthinkable in the United States. By the twenty-first century, animal welfare groups were almost universally opposed to the use of exotic animals in circuses and most definitely against the private ownership of such creatures. Horse slaughter has become an emotional issue, with the Humane Society of the United States and other organizations working to make it more difficult.

Even after accounting for the expenses of labor and of hauling stock in the zoo truck, the dressed meat cost only 30 cents a pound; the savings would have been greater if the Park Department had allowed the zoo to keep the proceeds of the sale of the animal hides. Unfortunately, the cost savings realized by the slaughterhouse did nothing to fundamentally alter the zoo's physical reality: it was still a second-rate facility, and only an infusion of New Deal funds would change that. Such an arrangement with the local humane society or SPCA was not uncommon across the nation. Still, Knudson's relationship with Washington's humane societies was not entirely pleasant, as revealed in the fall of 1935 when a small movement to investigate the care of animals at the zoo began with a complaint from Dr. R. D. Lelian to the Park Board and the City Council.[48]

1935—Attack on the Zoo

Lelian, a non-resident "naturalist," detailed his dissatisfaction with "cruel neglect" of animals at the zoo, which compared unfavorably to San Francisco's. Specifically, he charged that because raccoons were exhibited without access to drinking water one had recently died, that other animals languished in "bad, cramped quarters," and that the "condition of your animals and birds in many cases show improper feeding," particularly a lack of fruits and vegetables. Unfamiliar with Knudson and his history with the Park Board, Lelian blamed these problems on "inhumane, inefficient management," although he conceded that the zoo manager might "be new and inexperienced." Lelian wished that charges of cruelty could be brought against the manager, but in lieu of that he recommended that the city transfer its animals to New York City's humane Central Park Zoo, which Lelian mistakenly believed was managed by Al Smith.[49]

Knudson could take complaints about the condition of the zoo's quarters, which was largely beyond his control, but attacks on his integrity and qualifications were too much. In a newspaper story Knudson disputed Lelian's accusation, insisting that no raccoons had died there for several years. The only recent death was a tubercular monkey. "I don't know who this Dr. Lehman [sic] is," said Knudson, but "if he can show proof that any of our animals are mistreated or neglected he must be a magician. We keep a complete record of all deaths and the books are open to any one for inspection." He offered that the "keepers and myself are great lovers of animals and if any visitors notice anything that they think should be remedied we will be glad to cooperate to the fullest extent."[50] He might well have added that the zoo's truck farm and steady supply of surplus produce from local grocers meant that the animals enjoyed tens of thousands of greens each year. Moreover, Lelian's positive remarks about the humane condition of the San Francisco and New York zoos simply confirmed what Knudson had long argued: cities that used New Deal program wisely could create huge improvements in the welfare of their zoo animals.

Knudson's public response, however, simply generated more criticism of the zoo. One citizen wrote the city council to substantiate Lelian's charges and add his own: that the zoo kept some animals "in black underground places." Regarding the director, this citizen said that he and others "don't think Dr. Knudeson [sic] so considerate of all animals as he is trying to make out."[51]

The Board of Park Commissioners dutifully investigated the charges and responded to the city council in much the same way that it responded to Knudson when he asked for new zoo buildings. In the board's opinion, "the conditions complained of do not exist." While it was true that some of the cages were "rather cramped," they were no worse than at other zoos, and making them more spacious would "necessitate the expenditure of a considerable sum of money."[52] The board, notably, did not seize this opportunity to request such a sum in the next city budget.

The State Humane Society proved unwilling to let the board off the hook with its poverty plea. In the final days of 1935, it formally requested the board to address the inhumane conditions described by Lelian and others. Seemingly careful to avoid criticizing Knudson, the society president,

Redick McKee, identified "obtuse and indifferent Park Boards and City Councils" for their "insistence on confining" some animals in "black, damp dungeons" under one section of the zoo's cages. Here, in "deadly holes provided by our Park Board," the poor creatures paced "back and forth, back and forth, in their misery" while visitors made "commiserative comments." The problem was a political one, as was its solution, and McKee warned that the society was "determined to agitate this matter until this blot upon the humanity of Seattle's citizens is once and for all removed." In January 1936 McKee brought a group of animal welfare activists from his group and from the King County Humane Society to a Park Board meeting to suggest "curtailment of the Zoo" in light of the "limited funds available" for maintenance. Another citizen at the meeting submitted a written proposal for specific steps the Board could take to improve the zoo. For years, Knudson had been making similar complaints about funding and its effects on the animals' welfare, but no doubt still stung by the Lelian episode, he characterized the Humane Society appearance as a "publicity stunt," and he argued that in the face of "interference" from unnamed park department heads, he had "tried at all times to make conditions better."[53]

However unfair aspects of these complaints were, they were not so much different from Knudson's own assessment of the zoo, and they echoed his own concerns about animal welfare in general. Throughout his career, he too complained about outdated and inadequate exhibits. In nearly every piece of official correspondence, Knudson worked in some reference to the necessity of treating animals humanely, and he was quick to condemn others for mistreatment of animals. For example, in a report on the zoo's air raid defense plans, he ended by noting that the war had likely depleted the ranks of experienced animal men and he urged the city to "compel irresponsible persons, traveling shows, etc., to comply with basic humanitarian principles."[54] In an odd way, the Lelian charges had come at a good time. Works Progress Administration funds were still available, and the city could easily have seized on the welfare controversy to justify a full-scale redesign of the zoo. One particular resident of the zoo, Tusko the elephant, could have benefited from New Deal money.

Tusko

As the Depression deepened, a sign of hope arrived with the election of a new president who promised to right the nation's crippled economy. For Knudson, the period between FDR's election and his legendary first 100 days was an exciting one, but not because of expectations raised by the construction spree funded by the FERA and PWA. Rather, Knudson was busy trying to properly house and care for an unexpected gift from the mayor: America's largest living Indian elephant, Tusko. If all went well, Tusko's arrival promised to spur the city into building at least a proper elephant house, since federal funds were now available for just such a purpose.

Zoo directors frequently boarded circus animals as favors or to generate a little income, and it was common for circuses to donate older performing animals to zoos or even to abandon collections when times were tough. The great circus elephant Tusko, however, came to Seattle's zoo in an atypical way.[55]

Seattle Mayor John Dore, no great friend of the zoo, nonetheless found that when it came to protecting the welfare of animals, a zoo could be better than a circus. The elephant was a central attraction of circuses of this era. It was not unusual for a circus to have as many as ten elephants, and the Ringling had over 30. Elephants were popular in part because they were large, but they could also be dangerous. Many a "rogue" male elephant, one that could not be handled and had maimed or killed, ended up condemned to death by local authorities or sent to a zoo.[56] The latter was the fate of Tusko, the most famous rogue of the decade.

Tusko began his circus career with the more ordinary name of Ned, but as he grew and became more unmanageable, he was rechristened. Although Tusko had not killed anyone, he had knocked down enough people and walls to develop a bad reputation. Sold from circus to circus, his price dropped to the point where Bayard "Sleepy" Gray purchased him for one dollar. Teaming up with 21-year-old elephant keeper George "Slim" Lewis, Gray and another circus hand took Tusko on a one-elephant tour of the Pacific Northwest. They barely made it out of Portland, where Tusko tore down the building in which he was being exhibited. Down and out on his luck, Sleepy unwittingly signed over a controlling interest in Tusko to H. C. Barber, who promised to get the show on the road again. Tusko's reputation for violence, however, was headline news, and city official after city official refused to grant a permit for his exhibition. After languishing for a while halfway between Tacoma and Seattle, Slim finally got Tusko moving toward an expected payday in Seattle. Tusko's trailer, unfortunately, was not up to the task, spilling the large beast at the intersection of Marginal Way and Spokane Street in Seattle. With no license to exhibit Tusko, Slim and his companions tried to keep ahead of city officials, moving around the city for the next month or so. But it was hard to hide an elephant that stood 10 feet two inches and weighed a little over seven tons, especially with the newspapers covering stunts such as Tusko knocking over a house for the wrecking company Matheny & Bacon. As the money ran low and winter approached, Slim, the Humane Society, and Mayor Dore began to worry about Tusko's fate.[57]

In October of 1932, Dore ordered the confiscation of Tusko. To ensure that the "poor beast" would receive proper care, he was placed temporarily in the zoo, where Slim was hired as his keeper at the rate of $3.25 per day. Tusko lived up to his reputation, and within two months, he had attacked a zoo employee, prompting Lewis to post a notice "that no one except himself should go in [Tusko's] room."[58] Difficult to handle, expensive to keep, yet spectacularly large, Tusko created a real dilemma for the zoo, the city, and the Humane Society as they wrangled over how best to care for the massive beast. Whether Mayor Dore had given the zoo a gift or a burden was not immediately clear.

The Park Board, already stingy in its zoo appropriations, recognized that there would be additional costs associated with Tusko. A month before Tusko was seized by the city, the King County Humane Society asked the board to purchase Tusko outright. In late September, the board went on record as favoring the acquisition of Tusko, but only if given free title and a "public subscription to construct a building at the Zoo to house the animal." Humane Society representatives appeared at several more board meetings to advocate the zoo's acquisition of Tusko and support plans to build a permanent enclosure for him.[59] The board clearly wanted to avoid major expenses, but from an animal welfare perspective, it seemed that the zoo would provide more humane care than Tusko was currently receiving.

Woodland Park Zoo Director Gus Knudson also had reservations about taking on Tusko as a permanent resident precisely because of the animal welfare issues involved. Tusko was well-known to people in the zoo and circus businesses, and he had changed hands frequently over the past decade due to his unmanageable disposition. Rumors about him had grown, and he was wrongly accused of having killed two men and maimed many others. Knudson himself had once been called upon to remove 13 bullets fired into Tusko during one of his rampages. The Woodland Park Zoo's elephant house was barely adequate for its present resident, Wide Awake, and Knudson had been seeking and making minor improvements for years. Clearly, the present structure was far too small for a bull elephant of Tusko's size. Moreover, Knudson took great pride in the fact that Wide Awake was never chained, a sign of the good care and relative freedom that she received. By contrast, Tusko was kept restrained at all times with an elaborate system: all four legs were cross-hobbled, he wore a chain-basket on his head, his tusks were chained to his forelegs to limit his head movement, and a long chain trailed behind so that keepers could quickly secure him to a tree or telephone pole if he became dangerous.[60] His presence in the zoo would not convey the concern for animal welfare that Knudson was working so hard to cultivate.

On the other hand, it quickly became apparent that Tusko was both enormous and popular, drawing huge crowds to the zoo. Knudson called him "by far the greatest attraction our zoo has ever had," and crowds of up to 30,000 a day gathered to see him. The Park Board, however, vacillated on how to properly house the great animal. A temporary structure built of 12-by-24 inch beams was estimated to cost $7,000—too much for the stingy board to approve. Instead, the board authorized $850 to "remodel" the existing elephant house to accommodate Tusko, who was kept under a large tent in the meantime. The board continued to explore plans for a more permanent structure. Briefly, in Feburary 1933, it appeared the Park Board would support a plan to convert the archway under Phinney Avenue (which crossed zoo property) into a new home for Tusko, large and strong enough that his chains could be removed, allowing the public to see in "perfect safety" this magnificent animal unchained. Twenty adjacent property owners, however, petitioned against this plan, claiming that it would be "unsafe" for children and "the odor from this animal would be unbearable." Another group, the North End Clubs, also objected to housing Tusko in the underpass. Apparently bowing to this pressure, the Park Board went back to considering Knudson's requests for further improvements to the existing elephant house in the form of skylights and a radiator.[61]

But, as Knudson feared, Tusko created problems and attracted criticisms too. By April of 1933, Tusko was in his mating season and "on a rampage," tearing out his chains and beating his drinking tub into pieces. For safety reasons, Tusko remained chained, which was the "subject of much criticism" and led the Park Board to place a sign outside Tusko's enclosure explaining the "necessity" for this form of restraint. To counteract charges that Tusko was being mistreated, Knudson emphasized that on warm days he could be found outside "enjoying the sunshine."[62]

To permanently protect Tusko, the city had to find some way to gain legal title to the animal. Tusko's owner, H. C. Barber, understandably was reluctant to sign away such a famous animal without substantial compensation. Through his attorney, A.H. Henry, he threatened in January 1933 to take Tusko to Chicago if the zoo failed to come up with the money for the title to his elephant by March. Henry reminded the commissioners that Barber still owned the elephant even though it was in the custody of the city. Initially the Park Board made it clear that they did not have the funds to purchase or house the elephant. Portraying the city as the aggrieved party, the Board of Commissioners, under the leadership of President Barney Lustig, argued that the city had gone to considerable expense to take care of Tusko and resolved that "title to the elephant Tusko will not be relinquished until the Park Department has been reimbursed for all expenses incurred to date in connection with the care of this elephant."[63]

Henry suggested that the board "authorize Radio Station K.J.R. to raise funds by public subscription to purchase Tusko and donate [Barber's] elephant to the city." The radio station KJR agreed to conduct the campaign to raise the money to acquire Tusko for the zoo, as did N. F. Storm, the manager of radio station KOL. He proposed reducing expenses by having civic group representatives give the radio talks designed to raise the campaign funds. Ultimately the zoo and Park Board hoped to raise enough money to pay Barber for the title to Tusko and purchase a house for him. At most Barber was to get $3,000 and the rest of the money would go to the Park Commission. The fundraising campaign was on. Once it became clear that the fund drive might raise $8,000, some of which would go toward Tusko's building, Knudson became an active supporter, and a circus wagon to solicit donations was set up in front of Tusko's enclosure. Unfortunately, zoo visitors, gripped by the Depression, had "little money to give."[64]

Barber decided that he could raise more money for the Tusko fund by taking him to paid events around the city. The board, however, disagreed with him and probably worried about the safety of having Tusko moved around the city for fundraising purposes. Barber and his attorney also claimed to believe that the Park Commission was failing to improve Tusko's quarters fast enough. Sadly, on June 6, Tusko fell ill. Knudson treated him for several days with spirits of nitrate, sodium salicylate, and Barbados aloes, but Tusko died suddenly on Saturday, June 10, 1933, from a blood clot in his heart. On June 14th, Barber appeared before the park commissioners and asked for the body of Tusko. The commissioners agreed to his request, but stipulated that the meager $68.13 raised by the Tusko campaign should pay off existing debts on the fundraising drive, including some outstanding printing debts.[65]

Tusko's death brought predictable charges of neglect against Knudson and the zoo. In his defense, Knudson referred to Tusko's sad history as an abused and exploited performing elephant:

> I have observed this elephant for a good many years and have treated him at different times while he was owned by the Al G. Barnes' circus. Since leaving the circus, he has been used continually as a racket. The men who have handled him have used him for the same purpose. A threat to leave would bring higher wages. I have been charged in different ways for neglecting this animal while in our possession, have been accused of many things, but there is no foundation for any of these accusations. It is only loose talk. The elephant was better cared for while here than any time during his life. I have at all times refused to stand for racketeering.[66]

Tusko's death did not relieve the giant from legal wrangling and indignity. Barber, who still held legal title, arranged to have a tanner skin the body and prepare the bones and tusks for exhibition. Bankrupt and in debt, Barber apparently hoped to profit or at least break even from the sale or trade of Tusko's remains.[67] Such post-mortem treatment of a zoo animal generated little comment in the 1930s, but then the presence of death and loss at the zoo was much higher, caused in part by the economic exigencies of the Depression.

New Deal— CWA

Distracted by Tusko and by a Park Board that sought constantly to pare back the zoo, Knudson nonetheless began to learn about how other cities were using New Deal funds to transform their zoos, and his dreams of a possible future became more vivid. Where the board saw only economic crisis and belt-tightening, Knudson recognized opportunity. Specifically, Knudson hoped for barless enclosures and other naturalistic exhibits similar to those being built with New Deal funds elsewhere. Appealing to the Park Board's civic pride and concern for animal welfare, he used his monthly reports to offer glowing praise of what other cities were doing for their zoos. He passed on the comment of a visitor who had seen the Tulsa, Oklahoma, and San Antonio, Texas, zoos' new bear grottos and monkey islands (built with New Deal funds). This visitor praised their "natural beauty" and "how happy and contented the animals seemed to be." A few months later he reiterated the contrast between Seattle's zoo and those in other parts of the nation. One visitor "was quite enthusiastic" about the layout of the new Brookfield Zoo (Chicago); Knudson hinted that Seattle had never made any "plan for a *real* zoological park" of the kind enjoyed by Chicagoans. A former zoo employee returned from Texas and Oklahoma gushing with enthusiasm about the barless bear cages, monkey islands, and other "outstanding" exhibits in the San Antonio and Tulsa zoos. Knudson wrote longingly about how "contented" the Brookfield Zoo's polar bears looked in their new naturalistic enclosure. It was frustrating to see other zoos using New Deal funds to build such exhibits, and Knudson repeatedly tried to appeal to the Park Board's sense of competitive civic pride. In 1935, he stressed that the American Association of Zoological Parks and Aquariums annual meeting would be held in Cincinnati in 1936, where the parks had been improved "to a great extent" and the zoo was slated for still further construction, including a new reptile house. It hardly needed to be added that the nation's zoological directors would find nothing so impressive in Seattle.[68]

Knudson's constant hints resulted in a few new exhibits, many repairs of existing structures, but no major redesign of the zoo as in places such as Toledo, New York, New Orleans, and other cities. Compared to other cities, Seattle made only modest use of federal funds. Knudson surely felt frustrated by the fact that at the Depression's onset other cities had annual zoo budgets quite similar to or even smaller than Seattle's, yet managed to scrape together one-time outlays for building materials to take advantage of the generous federal matches. In Seattle, however, the CWA and WPA achieved only small improvements to the existing zoo, limited by the city's tiny allocations for materials.

But why would the city ask for so little money for animals? The answer lies with the city's mayor. Seattle's Mayor John Dore was a "pugnacious and demagogic" fiscal conservative who held his position from 1932 to 1938. He publicly expressed his desire to get rid of the Woodland Park Zoo altogether on the grounds that the public money was better spent on poor people who were starving and because he believed that the animals should be free. In contrast to the Park Board who planned to loan the animals, Dore believed that they should "reduce gradually the population of the Woodland Park zoo [sic] in anticipation of its elimination." He asserted that he never liked the idea of keeping animals "which were intended by nature to be free."[69] Zoos were a "relic of the dark ages." He argued that the money they saved by getting rid of the animals would free $1,000 a week that "could be put to far greater use by feeding hungry mouths and giving employment to Seattle families." Closing the zoo would be "both humane and economical."[70] Dore's facts and logic were shaky. The zoo's actual budget was about half of what he claimed, and closing the zoo would not "free" the animals but simply transfer them to different owners. Nonetheless, his political threat was hardly idle. Spokane, Washington, had decided to eliminate its zoo and had offered Seattle several of its animals.

While cities such as New York City seized the availability of CWA funds to begin massive new projects, Seattle used the program tentatively. CWA crews did finish a "zoo farm" to showcase domestic animals and acquaint city children with an unfamiliar way of life, but this low-cost project had actually been started in 1933 with contract labor. CWA workers also began construction on a naturalistic beaver pool with rocks, ponds, and waterfalls. Overall, though, the CWA did little because the Park Board — struggling with a budget that had been slashed 59 percent from its 1931 level — provided meager funds for materials, authorizing only $4,000 (from a state grant) in 1934. As a result, CWA workers could only make repairs to existing structures and walkways — painting, plastering, paving, fencing, and so forth. Still, Knudson tended to explain this work as done for the welfare of the animals. Other repairs addressed rat infestation, sanitation, and animal comfort and safety problems. For example, open wiring threatened electrical shock, and exposed beams and posts led gnawing camels to get slivers in their mouths.[71]

Such repairs were of high priority in 1934 because the nascent American Association of Zoological Parks and Aquariums planned to meet in Portland the following year, and Knudson was anxious to "be able to present nice looking buildings" to the conference attendees who would undoubtedly visit Seattle. Unfortunately for Knudson, he faced a stingy mayor and Park Board lacking the ability and imagination to commit the resources that would have allowed the zoo to make a major redesign for the welfare of its animals. Recognizing the political reality, Knudson prepared a list of 30 suggested zoo improvements for 1934, in rank order, with new, modern exhibits (a monkey island "to keep monkeys in open air," naturalistic big cat and bear grottoes "as are found in eastern zoos," a seal rookery, a new bird house) at the very end of his list.[72] These were the improvements that he most desired but was least likely to get. They were also the improvements that would have most benefited the welfare of the zoo's animals, moving them from cramped, barred, decrepit enclosures to open spaces.

To emphasize his frustration with the slow pace of zoo improvements, Knudson ended his 1934 annual report to the Park Board with a one-page memo titled "Special: For Your Information." Here he wrote:

> It is a proven fact all over the world that where there are attractive animals in any park the people are drawn to those institutions. It is not the fault of the animal that it is placed in captivity in cramped and unsightly cages with bad atmosphere, but the fault of man, who is superior in intelligence and who makes these conditions either good or bad for the captive animals. All up-to-date zoological parks are being constructed to make a more natural habitat for these animals placed in the parks. When visitors come to the parks everything looks as close to nature as it can be imitated, and instead of sympathizing with the animals they show enthusiasm. There have been a good many zoological parks established all over the country ... in the last year or so, when the authorities have had the opportunity of securing aid, both financially and in labor, from the government.[73]

In this missive, Knudson identified himself with the animals. Surely it was not *his* fault that Seattle's zoo animals lived in antiquated, inadequate cages. He suffered with the animals. No, it was *man's* fault, and the men in question were the politicians who devoted a measly 7.1 percent (about $31,000) of the annual park budget to the zoo, just enough for maintenance, but not enough to "make better conditions for these animals which have been given to us by patriotic citizens." His analysis pitted the politicians against the interests of both the animals and the public, who, by implication, desired "their" animals to be exhibited as humanely as possible. Elsewhere in the country "progress" was being made, thanks to "the help of the CWA and the relief workers," but in Seattle the zoo was receiving "no help."[74]

A year later in 1935, Knudson was even more specific about how little Seattle had utilized New Deal resources. He pointed out that "all through the east" federal funds had given zoos "bear and lion grottoes, monkey islands, swan lakes, amphitheatres, etc." But, "when all of this [WPA] money has been spent, like it was under the CWA, we will have nothing to show of any importance." Worse, some of the CWA labor had actually been directed to cut down Woodland Park's "natural growth," which "was no improvement, but rather destruction. The idea in constructing a park is to preserve nature rather than destroy it." Similarly, a zoo ought to protect animals, and so Knudson pleaded once again for "larger places" such as grottoes and monkey islands, "where specimens can be kept out in the open in this mild climate."[75] Knudson's consistency in stressing the welfare benefits of these new exhibition designs reminds us that such designs were not sought only or even primarily for their novelty or to create a superficial illusion of naturalistic habitat. Rather, zoo professionals believed that animals would be healthier in such exhibits, and the health of animals mattered greatly.

Knudson's complaints about the lack of support from the Park Board were certainly justified, but they had as much to do with the administrative structure of the board as with any intentional neglect. The unpaid five-member board was appointed by the mayor and served at his will. The board had jurisdiction over 44 parks, 42 playgrounds, 70 miscellaneous properties, ten bathing beaches, two golf courses, an art museum, and other recreational facilities. Yet, the annual Park Department budget was created by the city council, and the board had no authority to levy taxes. More important, the board, and thus the Park Department, lacked real leadership, a single person such as Robert Moses with the vision and authority to plan systematically. Civil servants in the Park Department carved out their own areas, with the "chief gardener" reportedly having the most power, while the board members came and went. As a result, according to critics, the parks' "development and operation" were "left to whim and chance."[76] Small wonder that Knudson's pleas went largely unheeded. The board's inaction, unfortunately, condemned the zoo's animals to substandard facilities, leading to mounting complaints and even calls to close the zoo.

WPA Legacy—Too Little, Too Late

Seattle was very slow to tap WPA funds to develop its zoo, and it sometimes seemed to do so only after external prodding by citizens or the Humane Society, although even that pressure did not always produce immediate results. WPA workers did complete the beaver pool begun by the CWA, and they performed a great deal of re-painting, paving, fencing, and other maintenance work, but no new exhibits were built during the first few years of the program's existence. Then, in 1936, the State Humane Society suggested to the city council that the next budget should allow a modest $3,000 in materials to be matched by WPA labor to build a proper infirmary for the zoo. The society pointed out that Knudson's back-office room had become too small for the zoo's large collection, and any "properly managed" zoo should have an infirmary. The city, in turn, appealed to the Humane Society to put up the $3,000 itself, which it declined to do. It took two more budget years, but in 1939 Knudson managed to have an old Civilian Conservation Corps building moved to the zoo where it was converted into a small, "up-to-date" infirmary with an "operating table, connection for ultraviolet-ray light, wash basin, cabinets for apparatus to be used for blood-testing, acid cabinet," and a "sanitary concrete floor." He enthused that this facility might "stop the criticism of the Humane Society for leaving the animals, which are sick, out in view of the public." It certainly helped satisfy the Society's desire for the "proper care of animals kept imprisoned for the pleasure of the public."[77]

Finally, in late 1938, the Park Board authorized a small project that had been sought longingly by Knudson for years: a monkey island, complete with a small schoolhouse (district # 13). Monkey islands were a ubiquitous feature of New Deal zoos. The concept was simple and very popular: instead of exhibiting monkeys in small, dark cages, dig a moat, build an island, erect some interesting objects for climbing — fake trees, a schoolhouse, a sunken ship, a lighthouse — and then populate the island with monkeys. Rhesus monkeys were exceptionally inexpensive, costing $10–$15 apiece when many birds went for at least three times that much, and they were quite lively. The public could view the exhibit from all sides, and large crowds gathered to watch the antics.[78]

Knudson had formally proposed a monkey island project, with plans and cost estimates, in 1937, but the board was slow to respond. Then, in July of 1938, the city council was presented with a petition by "a few of the boys and girls" who enjoyed the zoo but wanted "to see the animals in better yards and cages." The boys and girls noted how much they had benefited from the CWA farm, and they hoped that the city could find its way toward supporting a monkey island, since monkeys (presumably like children) "need this outdoor recreation and sunshine." The typed petition listed nearly 400 children and the address of each. It is difficult to escape the conclusion that Knudson, the Humane Society, or both had a hand in the preparation of this politically effective petition. After years of ignoring Knudson's requests for a monkey island, the board finally authorized the Park Department to spend $3,389.37 on the project, with the remaining $1,563 to be borne by the WPA. Completed in 1940, the moated, 89 by 72-foot monkey island promised to significantly improve the living conditions for the monkeys by taking them out of their cages and allowing them some freedom in the open air. As Knudson said, the monkeys "were enjoying their freedom for the first time during their captivity." A letter to the editor praised the Park Board, the "first one which has had any conception of what a modern zoo ought to be." An editorial hoped that this "step toward a long-needed modernization of the zoo" would not be the last.[79]

As in other cities, the monkey island brought publicity to the zoo in the form of newspaper reports of the new addition. Through these reports we see the humor with which the media treated events at the zoo, but it is also clear that from the media's perspective, the welfare of the animals was not taken seriously. The island opened in August 1940 and was enormously entertaining to reporters who kept up a steady stream of columns for several days about the power struggles on the island. In one article the reporter noted that a long-simmering "civil war" was reignited when "Little Elmer" was "plopped in the middle of the moat." This happened after the monkeys, who the reporter called "the board of directors," met at the schoolhouse. According to the reporter, Knudson erected the school hoping that "an educational system might be adopted that would create a democracy instead of an anarchy among the monkeys." The reporter recalled that he "heard some snooper" on his side of the moat say, "who do they think they are, anyway? Those monkeys don't know anything about higher education."[80] As the reporter and other spectators watched, the monkeys began fighting in the school and sent Little Elmer flying out of the door and into the moat.

Many of the conflicts ensued from Knudson's practice of putting as many monkeys on the island as he could, which meant mixing different species together and seeing whether they could coexist. Sometimes even similar animals will not tolerate one another, but Knudson was experimenting to see which animals would coexist peacefully and which ones they needed to move. According to Knudson, "If a monkey has no courage, is a coward ... the other monkeys sense it and drive him off. We never know until we put an animal out there whether he will be permitted to stay." According to reporters, Knudson was unconcerned about the battles. He was proud of the island because it was an improvement from cages and a "step toward a long-needed modernization of the zoo." Monkeys received injuries serious enough to require medical attention, but reporters shared Knudson's view, portraying the monkeys' fighting as a natural result of being freed from the isolation of their formerly cramped quarters. Of course they would fight — they did not yet know or trust each other. They were acting just "like humans."[81]

To emphasize the human-like dimensions of the island's social structure, reporters satirized the monkeys' battles as political conflicts. The storyline began in June with the sensational headline "Insurgents Heave Reactionary King Into Moat, Crown Suffragist," and it continued through the summer as one after another monkey sought to dominate the island. Knudson played along happily, offering nonsensical political commentary on the mayhem. A "Chinese" monkey, Sing, assumed power early, but "seventeen Filipino monkeys" took "matters into their own hands and drove him into the moat."[82] The overthrow brought cheers and screams from the watching crowd, and one concerned visitor called the police to rescue the "drowning" monkey. After Knudson's "Machiavellian" plans to have either Nebuchadnezzar or Jocko take over failed, he pinned his hopes on an African monkey named Dodey, a "civic minded, even politically minded" fellow who, in a more advanced civilization, "might organize clubs." Sing, however, regained power briefly before losing it to Cheeko, "a feminist left-winger" assisted by a "bloc" of dissenters "calling themselves the Young Turks." Knudson admitted that Sing "was perhaps too conservative. The opposition charged that he was a reactionary—some even claimed a Fascist." One reporter saw Sing's shortcomings as technical rather than ideological: "a good monkey, massive and well built, but no manager.... There's no science in him, no desire for learning. He didn't even attempt to go in the schoolhouse. He only sat on top of it." Sing's administration was also "marked by grafting, cheating and crooked work on the island," according to Knudsen. "Democrats, Republicans, and Socialists all had it in for him." Whatever Sing's political shortcomings, his frequent fights resulted in injuries that required treatment in the zoo hospital.[83] Knudson had sought the monkey island for years as a way to improve the quality of the monkeys' lives. Perhaps the island did this, but reporters focused on the island's entertainment value.

Other New Deal improvements, however, proved to be minor and received dramatically less publicity. The board grudgingly agreed to allow Knudson to secure some workers through the National Youth Administration program. And in 1940, a "goat mountain" was added to the project list. So, at the onset of World War II, the monkeys and the goats finally had some room to roam. The seals were less lucky, confined to a pool that the U.S. Department of Fish and Wildlife declared was "too small." To address this problem, the board authorized a WPA mammal pool project in October of 1941. Unfortunately, the WPA was in its twilight, and Seattle had missed the opportunity taken by other zoos to make numerous and substantial improvements in animal welfare. One unsatisfied animal rights crusader, despairing about the zoo's substandard exhibits—so ironic during a world war for "*freedom* and *liberation*"—could not "help thinking that if any Japanese bombing is done to the city, it would be lucky for the animals if the Woodland Park Zoo got all the first devastating hits, and all their unfortunate lives ended." As usual, the Park Board assured this citizen that Humane Society inspections had pronounced the zoo as doing all that was possible for the comfort and care of the animals.[84]

In 1947, two months before he announced his retirement, Knudson wrote a scathing review of a Proposed Development Plan for the zoo commissioned by the Park Board. In his review he insisted that the "primary concern" in any zoo design had be "to construct facilities for the keeping of animals alive." Architectural beauty was "worthless" if it lacked "accommodations for the welfare of the inmates." The proposed plan, in his view, failed to meet this central purpose. In a deeper sense, he argued that the city had long failed to devote the resources necessary to build a truly humane zoo, instead allowing the zoo to operate "on a hand to mouth" and "temporary basis," keeping its animals "cornered and cramped in one small portion of the park." When he retired later that year, Knudson vented his rage at his zoo's conditions to the press, characterizing it as "a prison and a disgrace to the city," a view soon after seconded by the *Seattle Times*, which called the zoo a "civic eyesore." Knudson insisted that he had "spent the best years of his life fighting in the interests of his animals against park board 'policies.'" The central problem for Knudson was that the zoo was too small, only 23 acres, but he also claimed that he "wasn't allowed adequate funds to feed the animals," that the buildings were "poorly designed," park facilities were inadequate, and the park board failed to devote sustained attention to improving the zoo. "Politics" had consistently hurt the zoo. Too many cages were double-decked, too small, and lacked ventilation. The zoo smelled bad. The one humane exhibit that Knudson could point to was the monkey island, designed by himself and built by the WPA.[85] This, then, was the unfulfilled promise of the New Deal in the Woodland Park Zoo that illustrates how local politics could thwart the best efforts of a zoo director to use the New Deal to create a more humane institution. What is remarkable is that the Seattle experience did not happen more frequently. Indeed, in the context of the human poverty and suffering during the Depression, the new zoos looked palatial.

Directors of major zoos in the 1930s thought a great deal about animal welfare. However, they had to temper their desires for animal enclosures with the fiscal reality of the Depression. Although they wanted better enclosures for the animals, they clearly believed that animal welfare primarily meant working as closely as possible with animals, nurturing them as one would human children. The welfare was the relationship between humans and animals. This was a faith that was lost (at least publicly) by later generations of zoologists, animal behavioralists, and activists who abandoned their belief in humans' ability to coexist easily with animals in the face of animal exterminations around the planet, injuries to employees at zoos, and trouble with violent guests. The logic was that because developers and hunters around the world were killing animals to the point of extinction, no humans could be trusted to ensure their welfare. This was coupled with the scholarly trend that emphasized de-anthropomorphisizing animals. Here the idea was that animals were not like us, or at least we could not be certain that they were. So, any attempts at looking for human qualities like love were misguided efforts to impose our own nature upon the animals. In short, zoological directors, keepers, and curators were doing more harm than good by coddling baby monkeys and treating them like small children. In the 1930s, however, zoo directors believed that this was the best way to treat exotic animals well.

SIX

The Decline, Resurrection, and Legacy of New Deal Zoos

Then of course the war came along and there was no more money for anything.

— Lucile Mann[1]

The entry of the United States into the Second World War imposed new hardships on America's zoos and, by 1943, spelled the end of the New Deal. War preparedness, rationing, manpower shortages, and curtailed animal importations all affected zoos negatively. So too did declining political support for the New Deal and for zoos themselves. In the decades after the Depression, many once-modern New Deal zoos fell into disrepair or simply became outmoded. By the 1970s animal welfare activists and sympathetic politicians called for the closure of these zoos. In a political climate where taxpayer support for zoos was often indifferent or inadequate, a new funding solution emerged: privatization. By creating private-public partnerships, many zoos previously built or rescued by public funding from the New Deal now staked their resurrection on an infusion of private capital. But the task of rebuilding New Deal zoos was complicated by their architectural, artistic, and historical legacy. In short, these zoos had to be both modernized and preserved. The result is that the physical legacy of the New Deal is still here for the public to view. What few can see, however, is the deeper human and animal legacy created by the New Deal. Saving America's zoos during the 1930s ensured that a generation of creative and conservation-minded zoo professionals had places to work. Today's robust zoo landscape — with its science and conservation orientation — surely owes a great deal to the New Deal.

World War II

World War II threatened exotic animals in American zoos at least as much, if not more, than the Depression. [2] At times the dangers animals faced during the onset of the war were imagined as worried citizens believed they were under imminent attack from the Japanese. The supposed threat of air raids on coastal cities required zoos to develop emergency procedures to prevent the escape of dangerous animals. Keepers at Woodland Park Zoo received instructions about how to secure the animals in the case of a bombing attack. "Should the need arise," dangerous animals could be put down with one of the zoo's three firearms. In the week following Pearl Harbor, director Knudson warned his keepers to be on the lookout for "marauders" or "parachute troops" with machine guns. Even enemy spies were a potential danger, and he asked his employees to "keep a watchful eye" on suspicious characters who might try "poisoning any of the inmates" or setting fire to the buildings.[3]

War rationing, in contrast, had real effects on the animals, making it even more difficult for zoos to get various foodstuffs. Zoos reduced hoofed stock by feeding some to the big cats. Other cats went on diets. In a "patriotic" move, Central Park Zoo Director Harry Nimphius, like his counterpart in New York's other zoos, cut the horse meat rations per cat by as much as two thirds, noting that "they're all fat" anyway. The monkeys and apes had to forgo bananas in favor of less scarce sweet potatoes. The Audubon Zoo simply put 14,837 pounds of mule and horse meat into cold storage in case of a "mule meat shortage," although it had to get a waiver from the Ration Board to buy an adequate number of beef hearts. When citizens in Washington State suggested that farmers reserve all hay for cattle, Gus Knudson indignantly defended the patriotic value of feeding his zoo's elephant, Wide Awake, which was purchased by donations from Seattle's draftees when they were children. "How would they like it," he asked, "when they come home to learn we had shot their elephant?" The question was not rhetorical. The Omaha Zoo destroyed several lions in 1942 because of the meat shortage.[4]

The war's disruption of international trade also complicated zoos' ability to acquire exotic animals from overseas. The effects of the war in colonial Africa made it challenging to meet the USDA requirement that animals spend 60 days in quarantine in their native continent prior to importation. Europe, previously a "clearing house" for exotic animals, was effectively shut down. Al Vida wrote the Bronx Zoo from South Africa in 1940 to apologize for not being able to get all of the animals requested; the war had made collecting "very difficult," as "all the hunters" were in the army. By 1943, one major New York City animal dealer was importing one-third fewer animals a year because of the war. The shortage had no major impact on America's many smaller zoos, most of which bought few or no animals from big dealers anyway, but it posed a potential problem for the elite zoos whose reputation depended upon their exhibition of rare and exotic animals that could be obtained only overseas. Even before the United States entered the war, directors from America's leading zoos — the Bronx, St. Louis, Buffalo, Detroit, and Philadelphia — met to discuss possible solutions to likely importation restrictions. Among the most significant was the idea of transferring animals among American zoos to support a captive breeding program.[5] Several decades later, the American Association of Zoological Parks and Aquariums would make participation in such breeding programs (called Species Preservation Plans) mandatory for member zoos.

The war's most significant threat to zoo animals came from the disappearance of capital improvement funds, which meant that necessary construction projects were put on hold. Zoo exhibits tended to deteriorate rapidly, especially pre–Depression structures; the addition of new species really required new exhibit spaces; and new knowledge about animal care changed ideas about the best exhibit designs. For nearly a decade, New Deal funding had allowed America's zoos to not only expand but also maintain a relatively modern physical landscape. The political and economic pressures associated with World War II, however, revealed the extent to which the nation's zoos depended upon federal largess. Anti-New Deal congressmen managed to eliminate the WPA and other public works agencies, and some local governments were unable or unwilling to do more than just provide annual operating budgets for their "modern" zoos, virtually guaranteeing that those zoos would soon decline in quality.

The New Deal building boom had also created the illusion of robust political support for zoos. In fact, the construction costs had been borne largely by the federal government, and when the new zoos were completed, local governments returned to their pre–Depression spending habits, allocating enough to pay the employees and feed the animals but not much more. In Buffalo, just a few short years after the WPA left, the zoo's beautiful new wrought iron fence was rusting, the new buildings still had unpainted windowsills, and director Marlin Perkins had had enough. Recognizing that the city government was unlikely to ever give him "enough money to run the zoo properly," he moved to the directorship of Chicago's Lincoln Park Zoo, where the financial prospects were greener. Three years after the new Audubon Park Zoo opened, a New Orleans Zoological Society member lamented that "we are unable to finance" the purchase of animals to fill the cages or even "afford a veterinarian to treat ill animals."[6]

Zoos also faced manpower shortfalls. In part, the federal projects were doomed because the unemployment problem had been solved. Relief workers joined the military, were drafted, or found jobs in the war production industries. By 1941 many WPA workers were shifted to defense-related jobs and others found employment elsewhere. For park departments such as Seattle's that had come to rely on relief labor for basic maintenance work, let alone construction projects, the war had a negative impact. Other zoos temporarily lost skilled young keepers who were drafted or volunteered for service. The Woodland Park Zoo's talented elephant keeper Slim Lewis found employment with Boeing during the war. Staten Island temporarily lost the services of its reptile curator after he enlisted. The Audubon Zoo reported a shortage of skilled labor, which it alleviated somewhat with prison labor assigned, ironically, to paint the iron cages.[7]

The politics of war shifted spending priorities at both the national and local levels. During the Depression exotic animals in zoos were sometimes seen as more deserving of funds than poor people. When faced with the decision of whether to allocate funds to soldiers or exotic animals, however, politicians and bureaucrats chose soldiers — a group that has typically received a disproportionate number of government sponsored benefits not available to ordinary Americans, including housing and health care. When politicians were asked to fund soldiers or animals, the animals lost. In Dallas, the Park Department director argued against a post-war bond proposal to fund capital improvements because "You've got monkey islands for monkeys ... pastures for zebras ... something for the lions ... but not a dime for G.I. housing." In the absence of city or federal support, no improvements were made to the zoo between the end of the New Deal projects (1939) and 1956. One of the first projects was a renovation of the WPA monkey house to make it more appropriate for the two lowland gorillas that had been acquired recently.[8]

In this state of war readiness, exhibits declined physically. Robert Moses had been praised in a 1935 Adlermanic Committee investigation of work relief for using public works projects "to achieve something of lasting value to the taxpayer."[9] In one sense, his three zoos were "of lasting value" — architecturally attractive and decorated by artists of national stature. Moreover, Moses had used New Deal funds to assure that these physical spaces were more or less permanent. Yet, it turned out that the physical integrity of Moses' zoos, like other New Deal zoos around the nation, was *not* "lasting." Zoos wore out relatively quickly, requiring constant maintenance and frequent renovation.

The New Deal structures, considered "modern" in their time, looked cramped and inhumane by the standards of the 1980s. Writing retrospectively and with a bit of historical revisionism in 2001, New York City claimed that even after its redesign in 1934 the Central Park Zoo "remained a place which many found a squalid place [sic]." The noise, the odor caused by the large animals, and the "depressing" succession of "spare cages" made it clear by the

1970s that "this zoo had outlived its usefulness." With the exception of the sea lion pools and bear pits, most of the exhibits more closely resembled prison cells than anything from nature. The animals lived on cement floors in small, barred cages attached to the building exteriors. Some exhibits, such as the Prospect Park Zoo's freestanding, wrought-iron birdcages, had been emptied because zoo professionals had learned that animals suffered psychologically when surrounded on all sides by people. An official report on the Prospect Park Zoo also noted that its cages were far too small, even though they had been described in 1935 as "bigger" than Central Park's, and Robert Moses had praised their bar-less enclosures as late as 1970. Both zoos had become plagued with the same kinds of problems that Moses identified when he first looked at the Central Park Menagerie in 1934. Prospect Park's naturalistic bear enclosures, for example, were infested with rats, and its other buildings experienced periodical basement flooding which contaminated the animals' food and water supplies and compromised the buildings' heating and ventilation systems. Together, these problems had already led to violations of the Marine Mammal Protection Act and the Animal Welfare Act, and ruled out any possibility of the zoo being able to meet the Endangered Species Act provisions that would permit it to display endangered species.[10]

By the early 1980s, the two city-operated zoos had become a "source of grave concern to both the public and city officials" and were having a "negative impact" on their surrounding areas. Citing "declining physical conditions and disturbing operational deficiencies," the report described an agreement between Mayor Edward Koch and the New York Zoological Society (NYZS) in which the city would invest a one-time capital outlay to renovate the zoos and the NYZS would take over all management and operational responsibilities, with limited annual financial assistance from the city. This agreement recognized that modern zoos required specialized personnel that city civil service hiring practices and salary limits could not accommodate, as well as an educational mission that fit poorly with the Parks Department's recreational mandate.

The NYZS — itself established in the nineteenth century with an educational, conservation, and research mission — emphasized that remaking the zoos into educational and conservation facilities would be a central priority. The 1930s era of "recreation" as a valid public interest had been replaced with an education mantra. In assessing the Prospect Park Zoo, the NYZS concluded that its New Deal design rendered the zoo "practically useless from the standpoint of education." The Society proposed a redesign that would make the zoo more "intimate," giving young children particularly the ability to see and learn about animals in small exhibits and through hands-on activities. The plans also proposed exhibits and affiliated materials that would support the City Board of Education's curriculum in world habitats. Interestingly, these proposals reflected the educational philosophy articulated by the SIZS a half century earlier, when it promised that its zoo would "be a model for future zoological development." A few miles away, the Central Park Zoo would similarly get rid of all its larger animals (with the exception of the sea lions and a new polar bear exhibit), shrink the collection, and reorient itself towards a conservation mission. [11]

Inadequate maintenance budgets were only part of the problem. Equally serious was the fact that knowledge about the best exhibit techniques evolved rapidly in the decades after the Depression. What had been touted as "modern" exhibition designs in the 1930s were all too soon perceived as inadequate and old-fashioned. The Cincinnati Zoo's reptile house, described in 1942 as "one of the most modern buildings in the world," was converted to a bird house in 1953, and the reptiles were relocated. The Audubon Zoo, a mere 30-year-old, had become a "zoological ghetto" in which animals were housed in "ancient cells." A comprehensive review of the zoo in 1971 criticized its New Deal layout as old fashioned, emphasizing formal garden design over the needs of the animals. Admitting that the cage design standards were "probably adequate by the contemporary standards" of the 1930s, the reviewers pointed out that they violated new norms and the Animal Welfare Act (1970).[12]

Monkey islands, touted in the 1930s as major improvements in animal welfare, were abandoned or redesigned to meet new standards and expectations. Every time San Francisco zookeepers went through the monkey island access tunnel, they had to dodge feces and urine barrages from the resident monkeys; in 1994 the island was demolished. The Woodland Park Zoo's monkey island, which had so delighted crowds in the 1940s, was described by consultants in 1970 as "a dispirited display" that needed to be replaced "with a facility that fulfills requirements of the primates using it." It is an open question as to whether the island was really dispiriting, and it seems more likely that the entertaining power struggles of monkeys in the 1940s were now perceived as evidence of the primates' unhappiness. The Audubon Zoo's monkey island was converted, with some irony, into a beer garden where the human species could engage in its own alcohol-assisted antics.[13]

Accompanying the physical decline of the zoo infrastructure was a gradual loss of some of the features that had made the New Deal zoo such a lively, popular place. The relationship between people and animals became more distant as animal welfare groups pushed zoos to make changes in exhibit designs and practices. And zoo employees furthered the distance between humans and animals when they sued their employers over unsafe working conditions with animals. Zoo professionals initiated a few of these changes themselves during the 1930s. The barless enclosure, for example, protected animals both by getting them out into the open and by putting them out of reach of visitors. Animals were safer when visitors were too far away to poke, prod, or feed them, but the zoo-going experience became a little less dramatic as a result. As we have seen, however, zoos compensated with a range of other practices that closed the distance between human and animal. Circus-style performances by trained animals were a standard attraction at most zoos. Opportunities to pet or even ride animals were also common. Animals were given names, dressed up, and generally treated as though they had personalities. And many zoos encouraged young visitors to build close relationships with animals — wild and domestic — outside the zoo's gates. Zoos hosted children's pet parades. They gave advice about how to collect and keep a variety of wild animals. In short, zoo professionals sought to develop a generation of Americans who would love and care for wild animals as they did. By the 1970s, nearly all of these practices were on the way out at the nation's leading zoos. Liability issues, pressure from animal rights groups, and new ideas about the appropriate relationship between humans and animals doomed these practices. All the animals were left with were the small cages that were only part of the idea of animal welfare practiced in the 1930s.

The Resurrection

However solidly built and architecturally beautiful, the zoo buildings of the New Deal era were subject to the forces of deterioration, and by the 1970s citizens and politicians alike voiced cries for renovation. Ironically, many critics focused on the New Deal buildings, once held up as modern, sanitary, and humane but now perceived as antiquated, dilapidated, and cruelly small. By the early 1970s, animal welfare activists had won a major legislative victory in Congress with the passage of the Animal Welfare Act and its amendments, as had conservationists with the Endangered Species Act and the Marine Mammal Protection Act. These regulatory developments brought American zoos under the scrutiny of the federal government and of the activists themselves.[14] Across the nation, animal welfare groups and their members publicized and protested inhumane zoo conditions. Other citizens raised budget concerns, arguing that local taxpayers could not afford to maintain a zoo.

Because of changes in the broader political landscape, zoos became more expensive during the 1970s. Unfortunately, the 1970s was also a decade of declining local tax revenues, expanding local government services, and a population shift to the suburbs and away from where urban zoos were located. For zoos, particularly the larger ones located in major cities, these changes threatened to reduce city appropriations and gate revenues at a moment of critical need. By the mid 1970s, most municipal zoos required more than just annual maintenance budgets; they needed to expand in dramatic and costly ways. External pressure from the Animal Welfare Act, the Endangered Species Act, and the American Association of Zoological Parks and Aquariums (shortened later to the AZA) required that exotic animals in American zoos receive more humane treatment to receive accreditation. [15] How could zoos get the funding for better enclosures and veterinary treatment? They turned again to the government.

It appeared for a moment in 1974 that Congress might support a Federal Zoological and Aquarium Assistance Act, which would have provided direct grants and loans to zoos for a variety of purposes including training, methods of improving animal welfare, and capital improvement money.[16] The federal assistance program for zoos, however, never got out of subcommittee. There would be no New Deal for zoos in 1974 or thereafter. Instead, zoos appealed to the private sector.

William Mann's zoo was on the forefront of attempts to get public funds and it was an early model for the turn to private money. With the end of the New Deal, the zoo was once again suffering from lack of funds, with little relief in sight. To help the zoo, a group of private citizens formed a Friends of the Zoo organization. The Friends charged membership dues, organized volunteers to work at the zoo, and — most important — acquired contracts for several zoo concessions: balloon sales, a gift shop, and a train ride. The revenue raised from these concessions enabled the Friends to hire an architect to create a master plan for the zoo and, eventually, to support new building construction.[17] This was the model that zoo directors around the country had to turn to for funds as cities and the federal government refused to help.

The turn away from governments and to private donors swept the country. In his memoir, zoo director Terry Maple interprets Zoo Atlanta's rise from an embarrassing, substandard zoo in the late 1970s to a national leader in the 1980s as a lesson in the benefit of creating a "public/private partnership" where the public retained ownership of a zoo while a private nonprofit organization took over management responsibilities. In such a partnership, the zoo enjoyed continued public financial support as well as the flexibility to "manage themselves in businesslike ways" and attract private funds. Such partnerships were relatively rare in the 1930s, when less than a dozen of the nation's zoos were set up in this way. Since then, however, much has changed. At the start of the twenty-first century, 85 of the zoos accredited by the AZA were governed by a society or other nonprofit organization. The media typically referred to this trend towards nonprofit management as the "privatization" of zoos, but this label oversimplified the transformation of zoos in at least two ways. First, it implied that formerly public zoos were completely private, for-profit operations; in reality, almost no public zoos have become entirely private in this sense. Second, the term "privatization" evoked a range of debates about government's "efficiency"; for zoos, however, privatization typically occurred not because of a desire to "do more with less," but because of a need to increase revenues and spending. Thus, while the term "privatization" was unavoidable, we should understand that from zoo directors' perspective it meant a "public/private partnership" created to maximize a zoo's ability to raise revenue and professionalize its operations for the welfare of their animals.[18]

Advocates of privatization lauded the success of zoos and ignored the extent to which most were still public institutions. Fiscal conservatives made strong claims for the benefits of privatizing other city services based in part upon the model of privatizing zoos. One proponent of privatization, for example, argued that the privatization of the Pittsburgh Zoo and the National Aviary in Pittsburgh were models that other zoos and public institutions ought to follow. In both cases, proponents of privatization saw superior private leadership and efficiency. The group that took over the aviary "did not hire a city employee to run it ... [but rather] hired someone with vision," and while it once "received 83 percent of its support from tax dollars; today the share is 57 percent and shrinking," apparently with no ill effects to the birds. Contrary to the fears of critics, advocates argued, the costs have only increased slightly at the aviary, employees adjusted to life as private employees "without incident," and the public was "generally unconcerned about the legal technicalities" of the change in governance at the institution. In addition, the personnel, "at least at the executive level," could be "hired on the basis of merit rather than seniority." And, "most important for animal lovers," advocates of privatization believed that "both institutions have secured their futures against city budget cuts." In another pro-privatization argument, advocates argued that privatizing zoos would create more "cost-effective operations" as well as "more exotic animals" and "grander housing." Private management of zoos, they maintained, would "improve the wellbeing of animals, ensuring not only clean concession areas but clean zoo cages." And zoos would attract more private sector donations if they privatized because "these sources tend to donate more to private organizations than to government programs, over which donors have little control." And finally, privatizing offered "stability" because "donors could be confident that the large sums they contribute for long-term projects will be used properly."[19]

While think-tank and scholarly advocates tended to stress efficiency and a "do it alone" attitude as the reasons to support privatization, zoos themselves took a much more nuanced position. Yes, zoos sometimes cited efficiency and "freedom" from bureaucratic control among their reasons justifying privatization, but when one examines the various reports, formal agreements, and newspaper accounts associated with zoo privatization, one finds a different set of arguments stressed.[20]

First and always foremost in every argument for privatization was the sense of impending financial doom precipitated by ever-stingier local government. Zoological landscape architect Jon Coe captured the general sentiment when he characterized privatized zoos as those that because of diminished municipal support "had hit bottom and faced closure." More concretely, the San Francisco Zoo saw its zoo budget cut by 20 percent in 1992 ($800,000), the city of Pittsburgh cut zoo funding in 1991 to ease pressure on a $35 million deficit, the Lincoln Park Zoo's city budget had been frozen from 1988 to 1993, and in Seattle a "shrinking city budget" could not meet the zoo's necessary expansions. In all of these cases, the zoos foresaw a need for significantly more spending, and the cities desired to trim their support for zoos. Because the AZA considers budget stability in its accreditation reviews, it will sometimes highlight the danger of declining city support. In the case of San Francisco, the AZA "tabled" the zoo's accreditation process amid concerns that the city could not "meet funding needs for the zoo given their current budget crisis."[21]

Just as animal welfare legislation and accreditation standards created a pressing need for improved facilities, they also required zoos to establish and fill a host of professional job positions. To care for animals, a full-time veterinarian needed to be on staff, and to fulfill education and research missions entire new departments sometimes had to be developed. Accordingly, zoos wishing to privatize tended to emphasize that under private management they would be able to increase their number of professional zoo staff. Pittsburgh's privatization plan called for staff increases in "animal health and research," including the "addition of a Research Coordinator, a position currently funded by a grant." In San Francisco, a central goal of privatization was to "develop an integrated, trained, professional staff," including the establishment of a "formal Research Department." New professional staff included curators, a registrar (to "improve collection management and compliance with federal, state, and local regulatory requirements"), a horticultural consultant, and educators.[22]

In short, by the 1980s, zoo directors had turned away from the model that either the federal government or local taxpayers ought to pay the whole cost of exotic animals at their institutions. In general, zoos sought privatization as a solution to funding crises, just as in the Depression they had hoped that work relief programs might pull them out of their economic woes.

Privatization modified the meaning of "public" zoo, and the quest for revenue affected even zoos that did not privatize. In the 1930s, not a single publicly owned and managed zoo charged a regular admission fee, and even the privately managed zoos either had free days or nominal gate fees — at most 25 cents, or about a quarter of what a relief laborer earned in an hour. Today only a handful of zoos do not charge admission, and typical gate fees are proportionally much higher than in the 1930s, averaging about $10, or nearly twice minimum wage. As zoos began to implement admission charges, they justified them as necessary to provide the zoo with reliable income, and argued that they had no real impact on attendance figures. This defense, however, seemed inconsistent with the simultaneous claim that an admission fee reduced the amount of vandalism, and with data showing that visitors spent less on concessions on free days at paying zoos. Somebody was kept out by admission fees, and it seemed that those people were the poor and the delinquent.[23]

Zoos also began to see concessions as revenue generators, not as public conveniences. In 1940, the San Diego Zoo earned relatively little revenue from its concessions ($6,000 compared to $60,000 from gate receipts), which the director perceived as being provided as a necessary service. As Belle Benchley said, "If the desire for profit overcomes the desire for good service, the organization of the zoo is judged as harshly by that as by its care of the animals and grounds." By contrast, in 1970 the San Diego Zoo earned 56 percent of its annual revenue from concessions profits.[24]

Significantly, even zoos that retained public governing authority adopted the strategy of relying on an active zoological society to attract the corporate sponsorship necessary to make renovations possible and withstand changes in the political climate. As Los Angeles Zoo Director Manuel Mollinedo said in October of 2000, "I want to get to the point where we're bringing in a significant amount of money from corporations and foundations ... so I don't have to rely as heavily on city funding [because] politicians, as a group, tend to be mercurial." A little-recognized tax law helped change the current face of zoo funding by increasing private sponsorship of exhibits at all zoos and aquariums. In 1992 Congressman Ed Jenkins (D-GA) sponsored H.R. 5645 that clarified the fact that "tax-exempt organizations receiving qualified sponsorship payments in connection with qualified public events would be excluded from unrelated business income tax."[25] Translated, this meant that zoos and aquariums could capture corporate dollars for exhibits. The law passed with the help of old zoo friend John Breaux (D-LA) who sponsored similar legislation that year in the Senate. The result was a sea change at zoos and aquariums as corporations began giving in a way that increased their profile and marketed their product. Zoos around the country developed exhibits such as the "Cheetos Cheetahs," the "Fuji Komodo Dragons," and "Alberston's Koala Outback" (in Fort Worth, Texas) that boldly advertised their sponsors.

The AZA, moreover, urged its members to aggressively seek these corporate dollars. Robert Ramin, director of development and marketing for the AZA, informed members of the findings about corporate giving detailed by Craig Smith in a *Harvard Business Review* article entitled "The New Corporate Philanthropy." Smith argued that "until recently corporate America feared environmentalism," but that "the new corporate philanthropy seeks to support sustainable development that accommodates business." Citing this new corporate giving, Ramin suggested that zoo professionals "cement relationships with corporate leaders" in their region. Ramin also informed zoo professionals that "funds from corporations are, more often than not, intimately tied to the use of a logo, coupon, or some type of promotion that involves selling the corporation's goods or services" because polling researching had found that "many consumers decide which products to buy, and at which stores to shop, by looking at advertisements and promotions that show how much a company gives to charity." He noted that zoos and aquariums were "well placed as popular partners" because of their "demographic appeal and conservation message."[26] Thus, by the 1990s both public and privatized zoos were tapping into corporate funding sources.

Some zoos also turned toward amusement park style entertainment to bring in revenue. For example, the AZA president was the keynote speaker at the 2001 International Association of Amusement Parks and Attractions convention; he spoke to the benefits of developing closer relationships between the two associations. How? Amusement rides. At least two major zoos have opened amusement rides since the millennium: the Fort Worth (Texas) Zoo and the Indianapolis Zoo. Interestingly, the Indianapolis Zoo built its "Kombo" roller coaster to replace lost revenue after the AZA recommended that it close down its elephant rides. Zoos have also taken control of their food services, copyrighting kid's meals marketing plans with such catchy titles as "AniMeals," "Feast with the Beasts," "Primate Pizza," and "Safari Spaghetti" (Brookfield Zoo, Illinois). Louis Cain and Dennis Meritt find that between 1975 and 1989 private donations to public zoos (including society owned) declined as a percentage of the zoo budget, while revenues from commercial activities such as gift shops and restaurants increased.[27]

Although increasing revenue through gift shops, concessions, and rides helped animal welfare because it raised the money available for animal care, it also made zoos appear less like institutions dedicated to conservation education and more like amusement parks. Zoo employees worried about this trend, suggesting such commercializing initiatives were evidence that zoos were sacrificing their primary mission to the values of the marketplace. One zoo professional mused that by ridding "Western society" of zoos that are recreational, zoos could "function more like museums, art galleries or national parks, which charge nominal or no admission charges and have more substantial government support." Relieved from their "'commercial capacities' more resources could (ideally) be devoted to conservation, education, and research imperatives."[28] The reality was, however, that zoos were equally commercial in the past, generating some of their revenue not from roller coaster rides, but from camel and elephant rides, as well as trained animal performances. As a result of animal welfare legislation and lawsuits, zoos can no longer use their animals in these "commercial" ways. Thus, the "commercialization" of zoos should be understood as an attempt to gain revenue to pay for the better educated staff to take care of the animals and comply with the laws that maintain their health and well-being.

And then the 2008 recession hit the United States. There were early signs that private donations might not sustain zoos at the level that directors had hoped. Even in the mid–1990s, the Seattle Zoo Commission reminded the city that the Woodland Park Zoo would "require public operating support — in perpetuity," as well as "access to public funding in the form of municipal debt financing." In fact, several city zoos decided against privatization by transferring control to a county, thereby taking advantage of a larger tax base. And at least one of the few zoos that completely privatized regretted the change early on. By 2003, poor weather and declining private contributions made Zoo Atlanta disappointed that it gave up a permanent public subsidy, for example.[29] The 2008 recession, however, dramatically slashed zoo endowments and made it clear, as the 1920s had to zoo directors, that entirely relying on the generosity of private donors was problematic and highlighted the extent to which zoos were still funded by public money.

By early 2009 zoos were facing huge losses in both private endowments and public funding. The Bronx Zoo, for example, experienced a shortfall in philanthropic giving. The Lincoln Park Zoo faced a budget shortfall of more than $1 million as private dollars vanished. The 2008–2009 recession highlighted the downsides of corporate giving which are that it can vanish quickly and that corporations' central goal of selfpromotion means that they typically fund special programs and exhibits, not operations or maintenance. Animals that live at the zoo, however, need food, clean environments, and medical attention, all of the time. Even the glamorous special programs were easily abandoned by corporations when they faced their own financial meltdowns.[30]

Trying to cope with their own budget woes, states also threatened to cut back the remaining funding for zoos. In New York, Governor Paterson proposed eliminating the Bronx Zoo's state-funding in two years, which would force it to eliminate 30 staff positions. In California city council members in Los Angeles cut funding for a $42 million elephant exhibit, and the state funded North Carolina Zoo in Asheboro was denied $4 million for repairs and exhibits. Florida state lawmakers cut $2 million for manatee hospitals at Lowry Park Zoo and aquariums in the state. The argument for cutting funding given by Governor Paterson was that the state chose to spend its environmental dollars on environmental groups reintroducing wild animals. He also planned to cut the funds for other "living museums" including botanical gardens. The Coalition for Living Museums, which supports 112 state organizations, faced a "potential 55 percent cut in its budget in the 2008/2009 fiscal year- from $9 million to $4 million"[31]

Unlike during the Depression, the federal government did not step in to help zoos. Mayors around the country, however, pushed for funding for their zoos from the federal stimulus bill (known as the American Recovery and Reinvestment Act). The Roger Williams Park Zoo in Providence, Rhode Island, for example, asked for a $4.8 million polar bear exhibit. The zoo's director, Jack Mulvena, asked for the money, like his predecessors during the Depression, because it would "stimulate the economy in Providence." This was the bill's ostensible purpose, which was to "direct funding at projects that are primarily and clearly aimed at benefiting the economic conditions of communities and the public at large." As in St. Louis during the Depression, the Coalition for Living Museums pointed out that zoos did exactly that. The Bronx Zoo and New York Aquarium, to take two examples, "generated more than $289 million in economic activity" in 2008. Zoo advocates noted that, like during the Depression, zoos' attendance had increased during the recession as local residents were choosing what were now called "staycations" to save money. Attendance at the Syracuse Zoo "increased 4 percent to about 345,000 visitors in 2008" and the Buffalo Zoo earned "$1 million in membership revenues for the first time." Mayors around the country asked for money for zoos and other recreational activities on this basis in an 800-page document that Pete Sepp, the vice president of the National Taxpayers Union dismissed as pork, arguing that these projects did not meet the high quality infrastructure projects that the bill was designed to fund.[32]

In the face of these financial difficulties zoo directors struggle to maintain the welfare of their animals. Advocates for zoos point out that unlike other kinds of museums they are unable to simply close a wing or store an animal in the basement until the economy recovers. Their commitment to the care of their animals and the requirements of the AZA accreditation and major pieces of legislation mean that the animals come first. That typically requires downsizing the staff, closing the institution for extra months, and ending environmental education programs so that they can spend money on food for the animals. Alternatively, they have sent their animals to other zoos around the country that have the space and funds to care for them adequately. But, as in the past, zoo leaders have wondered aloud about whether "this is the message we want to send"—that "the state can't care for the animals in our zoos?" Or "at the garden are they just going to let the flowers die?"[33]

The Legacy

Building the New Deal zoo had been relatively simple. Tear down the old, dilapidated, antiquated structures and erect new ones. Rebuilding the New Deal zoo after the 1960s was more complex. By then the New Deal structures were old, dilapidated, and antiquated. Almost all were older, sometimes by a factor of four, than the exhibits they had replaced back in the 1930s. Most were in various states of disrepair. Few met the minimum standards for animal health and keeper safety. The seemingly obvious solution to the problems caused by these old exhibits was to tear them down and start fresh. This solution ran up against an unexpected legacy of the New Deal: these structures and their adornments had historical significance. Significantly, however, zoos continued to use New Deal structures for other purposes that indirectly benefited the welfare of animals — transforming the structures into educational centers, shops, and restaurants that helped pay for the increased numbers of keepers, curators, and veterinarians that local taxes, federal government money, and private donations would no longer fund.

Two of the most distinctive features of many New Deal buildings were their architectural beauty and artistic adornment. Many buildings were designed to reflect regional architectural styles and utilize local materials such as native stone. Others followed art deco design principles and nicely evoked the spirit of the era. Unemployed artists with the Federal Art Project (a division of the WPA), many of whom went on to have distinguished careers, painted murals or carved statues, bas reliefs, or keystones for zoo buildings. By the 1970s, art historians had begun documenting this enormous body of work, and public arts advocates were agitating to restore and preserve it. Post office and school murals were uncovered, exhibitions toured the country, and books were published. Before long, historical preservation commissions and arts advocates gave New Deal public art legal protection.

In New York City, a large number of significant artists had worked on the New Deal zoos. When a consultant researched the Prospect Park Zoo in 1980 to determine what the Parks Department should "consider for historic preservation should the need arise," he identified 28 separate items. Most prominent among these were the 19 reliefs and keystones carved by F.G.R. Roth, Emile Siebern, and Hunt Diederich on the entrance gateway, the lion house, the snack bar, the elephant house, and the monkey house. In addition, he identified eight murals on canvas painted by Allen Saalburg. "The rest of the zoo," he concluded, "can be plowed under for all I care." The difficulty of "plowing under" those buildings with the artworks in question was made apparent in an internal Parks Department memo a year later, which noted that these historic properties "may become factors in the New York

The Staten Island Zoo's 2007 remodel of its reptile wing included a recreation of reptile curator Carl Kauffeld's office (Staten Island Zoological Society).

Zoological Society take over agreement." Specifically, the art works fell under the jurisdiction of the Art Commission and the Landmarks Preservation Commission, both of which would have to "approve additions, deletions, relocation or alteration" of the identified works.[34] In other words, ironically, the historic significance of the New Deal structures was, in a legal sense, of higher priority than the welfare of the animals housed in those buildings.

The Central Park Zoo enjoyed a similarly rich artistic legacy. The old Arsenal building interior featured whimsical murals painted by Saalburg. Thus, the New York Zoological Society carefully incorporated this valuable art into the new and renovated buildings. Although all but four of the old buildings were demolished, two of F.G.R. Roth's Central Park Zoo sculptures were relocated, and 16 limestone bas reliefs by various artists were carefully removed and integrated into the new buildings.[35] Similar efforts preserved important New Deal art in the Prospect Park Zoo.

Architects for the Audubon Zoo also preserved some important New Deal art in their mid–1970s renovation. Constructed of brick, with slate roofs, the zoo buildings were relatively attractive, despite their fundamental inadequacy as exhibition spaces. A 1971 zoo improvement plan called for nearly every New Deal animal building to be renovated for a new, non-exhibition

In 2007 the Staten Island Zoo remodeled its New Deal reptile wing, renaming it the "Carl F. Kauffeld Hall of Reptiles" after the herpetologist who had played such a crucial role in the zoo's early history. Notice the tile mosaics; the original WPA exhibit murals were lost, but the remodel beautifully preserved the spirit of the New Deal (Staten Island Zoological Society).

use. The large hoofed stock building became a children's play area, its tower dovecotes sealed up to keep out the offensive pigeons. The old tropical birdhouse featured a beautiful brick relief of animals leaving Noah's ark carved by Guggenheim award-winning artist John McCrady; this structure became the new "Earth Lab," an environmental education center.[36] Another building with a similar McCrady relief was adapted to a new, non-animal use. Sadly, the site plan oriented traffic around the buildings in such a way that McCrady's artwork is largely out of sight. Only visitors who know where to look will appreciate this creative gem from the New Deal era.

In addition to the artwork in zoos, the structures themselves could be of historical significance. Some architectural movements received a tremendous boost from New Deal construction projects. The American Rustic Movement, for example, was particularly appealing because of its emphasis on the use of local materials to construct buildings, benches, and other facilities that blended into the natural environment. Parks throughout the United States featured benches, comfort stations, service buildings, and other structures built in this distinctive style. In Pueblo, Colorado, the city zoo featured five such structures, all built with native sandstone under the PWA and WPA: the Animal House, Monkey Island, Monkey Mountain, Bear Pits, and Tropical Bird House. For a modest city zoo, the New Deal enabled the creation of some relatively grand exhibits. The Monkey Island, for example, was anchored by a 40-foot tall lighthouse and decorated with a wrecked ship "named after Ray Talbot, the City Commissioner responsible for bringing many federal work programs to Pueblo during the Great Depression." The animal houses were adorned with 17 bas reliefs carved by WPA workers, as well as lion, bear, and gorilla sculptures. Other artistic features included a sculptural drinking fountain and a decorative pool graced by concrete seals. New Deal workers had given the city a zoo of unique architectural and artistic value, but by the 1990s all of these structures needed repair and some were no longer suitable for the exhibition of animals as originally designed. The Animal House, in fact, had deteriorated to the point where its safety for human occupants could not even be guaranteed. Yet, the Zoological Society Board of Directors was reluctant to tear down the building: "it seemed demolition would also destroy a very important part of Pueblo's history — the

The Staten Island Zoo remodel of the reptile wing retained the building's original exterior but opened up the interior exhibit space (Staten Island Zoological Society).

critical WPA projects that saved our community from destitution during the Great Depression." Instead, the board worked to get the several New Deal structures listed on the National Register of Historic Places as the Pueblo Zoo Historic District. So, in 1999 the zoo embarked on a renovation plan that would preserve the zoo's architectural heritage while adapting the old structures for modern uses. In 2007, the zoo opened a $2.9 million exhibit—Islands of Life—that incorporated the old Animal House and Monkey Island. Such restoration and renovation work cost much more than it would to have simply torn down the old structures and started over.[37]

In their restoration, New Deal zoos have once again become sources of local pride. Some are modest, like the Washington Park Zoo's charming rustic-style paths and buildings. Although small and not even accredited by the AZA, this zoo is on the National Register of Historic Places, a fact that the city emphasizes. In Toledo, Ohio, the zoo's New Deal buildings have been beautifully restored and adapted to new exhibition purposes. Informational signs, well-illustrated with old photographs, tell the story of how and why these structures were built during the Depression. This zoo grounds itself in that history. In Chicago, where exceptionally talented artists did much work under the Federal Art Program, the Brookfield Zoo has mounted a spectacular display of New Deal work done for the zoo.

However, the renovation of New Deal zoos has proceeded at a much slower pace than their original construction. As Mayor Edward Koch dryly observed when he announced plans for the new Central Park Zoo, the old one took "16 days to design and eight months to build. This zoo took two years to plan and will take six months to demolish and three years to build." The cost of modern zoo construction, borne entirely by local revenues, was enormous. Renovating the Prospect Park Zoo took nearly a decade and required $37 million. The Staten Island Zoo completed an award-winning remodeling of its reptile wing in 2007 at a price tag of $13 million.[38]

This history reminds us that a one-time infusion of federal funding during the 1930s did not address local government's long-term challenge of raising the revenues necessary to maintain and upgrade zoos. Harry Hopkins had promised that one of the great advantages of New Deal works projects in the parks was that their maintenance costs would be so low; he imagined that modest user fees would adequately fund their upkeep. More realistically, Moses had prophetically warned in 1940 that his park projects would incur large costs for future generations, costs that he urged taxpayers to accept. As he observed, every new project meant "running up the bill for personnel, maintenance and operation." Even if the federal government had borne most of the cost for the zoos and other park projects, local taxpayers would need to keep them "properly maintained." Accepting the "need for economy and the constitutional limits of taxation," he nonetheless insisted that even in 1940 New Yorkers were spending too little on their park system.[39] These concerns proved to be especially relevant to the city's zoos, which not only required a tremendous amount of regular maintenance but also came to need major remodeling to meet evolving animal welfare standards. The Prospect and Central Park Zoos, unfortunately, had been designed to exhibit large animals in small spaces, and since both zoos were run by a recreation-focused Park Department — rather than by an education-oriented zoological society — there was no internal pressure to change the zoos' basic mission. If the federal government had continued to fund local infrastructure projects after World War II, the fate of New York's zoos might have been different.

It may be a cliché to say that a society's treatment of animals is an indicator of its broader commitment to the welfare of its citizens, but the history of our nation's zoos suggests some truth in the claim. During the roaring twenties, some Americans prospered as the gap between the wealthy and the poor grew to historical extremes and the condition of America's zoos declined. After FDR's election, the welfare of people and zoo animals became literally entwined, with the unemployed set to work building more humane facilities for exotic animals at the nation's zoos. During the Depression the federal government and local citizens could easily have decided that exotic animals in zoos were among the least deserving of funding, but, on the whole, they chose to protect and support those animals. In the process they built up local communities by providing jobs to humans who were desperately in need of them, creating lasting structures that continue to be used to this day, and nurturing a generation of young zoo professionals who went on to be leaders in their field. However inadequate some of the structures were in retrospect, they literally saved the lives of animals that might well have been destroyed otherwise, and they gave young zookeepers a place to work. It is much more difficult to see this human legacy, but the New Deal built institutions, not just buildings. Within those institutions, young men and women learned the zoo business and developed a conservation ethic. Within a few decades, they were the leaders of the American Association of Zoological Parks and Aquariums, organizing the zoo community to survive another assault on zoos — this one ideological as well as economic. The modern zoo, although now funded largely by the private sector, surely owes much of its existence to the New Deal.

Chapter Notes

Preface

1. Pamela Henson, Lucile Quarry Mann Transcript, June–August 1977, Box 1 RU 9513, p. 174, 168, Smithsonian Institution Archives, hereafter referred to as SIA. Thomas Crosby, *A Zoo for All Seasons: The Smithsonian Animal World* (New York: W.W. Norton, 1979), 35.
2. Pamela Henson, Lucile Quarry Mann Transcript, 43, 37–39, SIA.
3. Pamela Henson, Lucile Quarry Mann Transcript, 36, SIA; "finest" quotation from William Mann to Mrs. Vera Schroeder, 22 Dec. 1939, Box 253, Folder 2, RU 74, SIA. Pamela Henson, Lucile Quarry Mann Transcript, 89, SIA.
4. See, for example, Vernon N. Kisling, Jr., ed., *Zoo and Aquarium History: Ancient Animal Collections to Zoological Gardens*; Elizabeth Hanson, *Animal Attractions: Nature on Display in American Zoos*; Ted Ligibel, *The Toledo Zoo's First 100 Years: A Century of Adventure*; Jesse Donahue and Erik Trump, *Political Animals: Public Art in American Zoos and Aquariums.*
5. Ken Kawata, "Cultural Icons, Paychecks, and Litter: Returning to the Roots of Zoos," *International Zoo News*, Vol. 49, No. 8 (2002), pp. 452–464. Quote from p. 452.
6. Ken Burns.
7. Pamela Henson, Lucile Quarry Mann Transcript, June–August 1977, Box 1 RU 9513, pp. 133–47, 158–59, 162, SIA.
8. Pamela Henson, Lucile Quarry Mann Transcript, 164, SIA.
9. Official WPA and PWA records are held at the Library of Congress's off-site location in Baltimore. These records, however, collapse funding information about zoos with all other information about parks, making it nearly impossible to get a clear sense of total funding for any particular zoo. Luckily, the individual zoos we researched often, although not always, had complete records of the amount of money provided by New Deal agencies for that institution. Nonetheless, we have not attempted to estimate the total amount of federal money spent on all zoos during this period.

Chapter 1

1. Interview in the *New York Herald-Tribune*, quoted in Robert Moses, *Public Works: A Dangerous Trade*, 3.
2. American Association of Zoological Parks and Aquariums, *Zoological Parks and Aquariums* (1932), 96–100.
3. L. C. Everard.
4. Figures from Everard. The numbers do not add up to 88 because some zoos did not report budget information. Judging from the other reported information, however, it is safe to assume that those latter zoos would fall into the bottom category.
5. Rich Buickerood, 1–9. "First class" quotation from Unknown to Mayor Sawnie R. Aldredge, 13 Jan. 1922, "Mayor" folder, Box 6, Park and Recreation Department General Subject Files, 1911–1970, Dallas Mu-

nicipal Archives, Office of the City Secretary, City of Dallas, Texas.

6. See Jesse Donahue and Erik Trump, *Political Animals: Public Art in American Zoos and Aquariums* (Lanham: Lexington Books, 2007) or William Bridges, *A Gathering of Animals: An Unconventional History of the New York Zoological Society*.

7. Raymond L. Ditmars and Lee S. Crandall, *Guide to the New York Zoological Park*, viii–xii.

8. Charles Haskins Townsend, 1–6. The claim about the aquarium's relative size is difficult to confirm.

9. Gary Zarr to Jonathan Kuhn, 28 Oct. 1987, "Central Park Zoo: History and Evolution," Zoos file, New York City Department of Parks and Recreation Library (hereafter NYC Parks Library]; Caroline A.A. Baker, 17–18, Central Park Zoo file, NYC Parks Library; Joan Scheier, *The Central Park Zoo*, 30–31.

10. *Brooklyn Parks Department Annual Report, 1894*, 59, Prospect Park Zoo file, NYC Parks Library; *Brooklyn Parks Department Annual Report, 1915*, 164–65, Prospect Park Zoo file, NYC Parks Library; Joan Scheier, *New York City Zoos and Aquariums* (Charleston, SC: Arcadia, 2005), 51–53.

11. David Ehrlinger, 62.

12. Wilbur L. Matthews, *History of the San Antonio Zoo*, 9–10, 14–17, 21, 26.

13. New Orleans Zoological Society letter to members, 30 March 1929, Box 56–15 Correspondence; APC meeting minutes, 21 Nov. 1928, Box 56–149 Minutes of APC 1927–1933; "Audubon Park" [press release], 26 May 1937, 56–142 Minutes of Audubon Park Commission. All in Audubon Park Commission Records, Earl K. Long Library, University of New Orleans (hereafter APCR).

14. Information taken from the Lake Superior Zoo, "Narrative History," <www.lszoo.org/history1.htm> (6 April 2007).

15. *Tulsa Zoo: The First Fifty Years*, 2–12.

16. Audubon Park Commission meeting minutes, 21 April 1933, 56–149 Minutes of APC 1927–1933, APCR.

17. Ed Bean to Edmund Heller, February 1935, quoted in Andrea Ross, 59.

18. Report of the Chairman of the Building Committee, 13 Jan. 1938, Grounds and Building Committee 1938, Box 90, Brookfield Zoo Archives.

19. Untitled editorial from *The Globe-Democrat*, 7 August 1931, St. Louis Zoo Records, 1910–1940, Roll 6, Western Historical Manuscript Collection, University of St. Louis–Missouri.

20. Gus Knudson, Report to Board of Park Commissioners, 1 May 1933, 3, folder 51/6, Woodland Park Zoo Monthly and Annual Reports 1933–1936, Don Sherwood Parks History Collection, Seattle Municipal Archives.

21. "Zoo's New Cheetah Ill," *New York Times*, 20 July 1931, ProQuest Historical Newspapers.

22. A review of the Audubon Zoo in 1971 criticized the practice of feeding donated French bread, concluding that its "only advantage is a reduction in the money spent for food at the expense of the animals' nutritional health." *Audubon Park Zoo Study: Part I—Zoo Improvement Plan* (New Orleans: Bureau of Governmental Research, 1971), 76.

23. For example, the PWA funded new building construction in the Saint Louis Zoo. St. Louis Zoological Garden Annual Report for 1939–1940, *The City Journal* (11 June 1940), 6, Office of the General Director Allyn Jenkins folder, Wildlife Conservation Society Archives.

24. For discussions of the PWA, see Bonnie Fox Schwartz, *The Civil Works Administration, 1933–1934: The Business of Emergency Employment in the New Deal*; Robert J. Leighninger, Jr.

25. *The Emergency Work Relief Program of the F.E.R.A., April 1, 1934—July 1, 1935*, 2–3, 5.

26. Harry L. Hopkins, *Report on Progress of the Works Program, October 15, 1936*, 106.

27. *Emergency Work Relief Program of the F.E.R.A*, 1.

28. Hopkins, *Report*, 107; Jason Scott Smith.

29. *Tulsa Zoo*, 10.

30. Superintendent's Report, 18 Dec. 1940, Annual Reports 1940–41, APCR.

31. William Mann to Mrs. Vera Schroeder, 22 Dec. 1939, Box 253, Folder 2, RU 74, Smithsonian Institution Archives.

32. Jacob Baker to Harry Hopkins, 31 Dec. 1935, reproduced in *The Emergency Work Relief Program of the F.E.R.A.*, n.p.

33. *Emergency Work Relief Program of the F.E.R.A*, 95.

34. Phoebe Cutler (p. 22) claims that 17

zoos were built entirely by the New Deal, but she gives no source for this figure. Still, the number is about right if one includes the zoos where relief programs *added* significant new structures.

35. Roger Conant, *A Field Guide to the Life and Times of Roger Conant*, 42.

36. La Guardia, quoted in H. Paul Jeffers, 177.

37. Robert Caro, 368.

38. Caro, 426.

39. Jeffers, 227, 231.

40. Caro, 368–69, 453.

41. Conant, 38–40.

42. Buickerood, *History of the Dallas Zoo*, 8.

43. David Ehrlinger, 65–69; Ohio Writer's Project, *Guide Book: The Cincinnati Zoo*, 5.

44. *Emergency Work Relief Program of the F.E.R.A.*, 95.

45. Details about the construction of the Brookfield Zoo taken from the monthly reports of the Grounds and Building Committee, 1934–1941, Brookfield Zoo Archives.

46. Buickerood, 8–11.

47. "Aquarium–Construction–General," Box 1, Park and Recreation Department General Subject Files, 1911–1970, Dallas Municipal Archives, Office of the City Secretary, City of Dallas, Texas.

48. Wilbur L. Matthews, *History of the San Antonio Zoo*, 26–29.

49. Mildred F. Heap, 14–15.

50. Robert Moses to Colonel Brehorn Somervell, 23 Oct. 1936, Box 97, Robert Moses Papers, Manuscripts and Archives Section, New York Public Library. The bulletin in question was Operating Procedure no. 0–6, 12 Oct. 1936.

51. A Taxpayer to Robert Maestri, 9 Aug. 1938, Mayor Robert Sidney Maestri Records 1936–1946, Correspondence 1937–1938, Microfilm Reel #406B, New Orleans Public Library.

52. Jeffers, 163.

53. See Jesse Donahue and Erik Trump, *Political Animals: Public Art in American Zoos and Aquariums* (Lanham: Lexington Books, 2007); William Bridges, *A Gathering of Animals: An Unconventional History of the New York Zoological Society*.

54. Caro, 458–60.

55. Editorials (1922 and 1926) quoted in Caroline A. A. Baker, 17, Central Park Zoo file, NYC Parks Library; *Annual Report, Department of Parks, Manhattan* (1929), 61–62, NYC Parks Library; *Annual Report of the Department of Parks, Borough of Manhattan* (1931), 82, NYC Parks Library; "Low Speed in Park Favored by Herrick," *New York Times*, 27 Sept. 1931, ProQuest Historical Newspapers.

56. Robert Moses, notes in response to a *New York Times* reporter's written questions about Aymar Embury, 26 March 1936, Robert Moses Papers, box 11, Park Book, Manuscripts and Archives Section, New York Public Library (hereafter RMP); Robert Moses, speech on occasion of presentation of honorary degree to Aymar Embury, 4 Feb. 1951, box 135, RMP; *The Report of the Department of Parks to August 1934*, p. 148, NYC Parks Library; "New Zoo Designed by CWA in 16 Days, *New York Times*, 27 Feb. 1934.

57. "New Zoo Designed by CWA in 16 Days, *New York Times*, 27 Feb. 1934; "Bricklayers Begin Work on Park Zoo," *New York Times*, no date, Central Park Zoo file, NYC Parks Library.

58. "New Zoo Opens," *New York Times*, 3 Dec. 1934, ProQuest Historical Newspapers; Robert Slayton, 378–79, 395; Robert Moses to Alfred E. Smith, 25 July 1935, box 11, Governor Smith, RMP.

59. "Moses Defends High Park Costs," *New York Times*, 20 Jan. 1940, ProQuest Historical Newspapers; Aymar Embury II to Robert Moses, 21 Aug. 1936, box 11, folder E, RMP; Robert Moses to Aymar Embury II, 27 Aug. 1936, box 11, folder E, RMP.

60. "Central Park Zoo to Get New Houses," *New York Times*, 20 Feb. 1934.

61. Untitled press release on the new Prospect Park Zoo, Department of Parks, 3 July 1935, Prospect Park Zoo file, NYC Parks Library.

62. *1894 Brooklyn Parks Department Annual Report*, p. 59, photocopy in Prospect Park Zoo file, NYC Parks Library; *1915 Parks Department Annual Report (Brooklyn)*, p. 164, Zoo file, NYC Parks Library; "Wants More Animals in Prospect Park Zoo," *New York Times*, 17 June 1923, ProQuest Historical Newspapers; *Minutes, Department of Parks Board of Commissioners* (1929), 153, 161, NYC Parks Library; *The Report of the Department of Parks to August 1934*, pp. 164–65, NYC Parks Library.

63. "Smith Now 'Agent' for Prospect

Zoo," *New York Times*, 10 May 1935, ProQuest Historical Newspapers; Robert Moses, notes in response to a *New York Times* reporter's written questions about Aymar Embury, 26 March 1936, box 11, folder E, RMP; "Civic Groups Tour Brooklyn Parks," *New York Times*, 25 July 1935, ProQuest Historical Newspapers; untitled press release on the new Prospect Park Zoo, Department of Parks, 3 July 1935, Prospect Park Zoo file, NYC Parks Library.

64. "600-Pound Pet Bear Kills a Cub, Claws 2 Others in Central Park," *New York Times*, 21 Sept. 1935, ProQuest Historical Newspapers.

65. See, for example, "Smith Visits Zoo, Feeds His Charges," *New York Times*, 18 April 1935, ProQuest Historical Newspapers.

66. "Central Park Zoo to Get New Houses," *New York Times*, 20 Feb. 1934, ProQuest Historical Newspapers.

67. "From Our Readers: Zoo Project Brings Echoes of General Satisfaction," *Staten Island Advance* (13 March 1934), Staten Island Zoo scrapbooks, Staten Island Zoo Archives (hereafter SIZA).

68. Alexander Crosby, "16 Obstacles Hold Back New Zoo," *Staten Island Advance* (22 Dec. 1933), clipping in Staten Island Zoo scrapbooks, SIZA.

69. Untitled press release (Department of Parks), 6 March 1934, Barrett Park/Zoo file, NYC Parks Library.

70. "Inspect Staten Island Zoo," *New York Times*, 25 June 1934; "Moses Informs Chamber He'll Push Park Projects," *Staten Island Advance*, 5 June 1935; "Park Plans to be Told Tomorrow," *Staten Island Advance*, 25 June 1935 [the zoo's final cost was $500,000]; "Finish the Zoo!," *Staten Island Advance*, June 1935; "A Jab for Mr. Moses," *Staten Island Advance*, 6 July 1935; "400 on Island WPA Jobs Quit Work in Protest," *Staten Island Advance*, 4 Oct. 1935 — all clippings in Staten Island Zoo scrapbook, SIZA.

71. "No Advertisement for Mr. Moses," *Staten Island Advance*, March 1936 [undated clipping], Box 97, RMP; Robert Moses to Harold J. O'Connell, 18 March 1936, Box 97, RMP; "New Barrett Park Zoo to Open Early in Summer," *Staten Island Advance*, 14 April 1936, Staten Island Zoo scrapbooks, SIZA.

72. "Staten Island Zoo to Open Wednesday," *New York Times*, 7 June 1936; "Staten Island Zoo to be Opened Today," *New York Times*, 7 June 1936, ProQuest Historical Newspapers.

73. L. Ronald Forman, 127–130.

74. Audubon Park Commission meeting minutes, 19 June 1935 and 17 Nov. 1937, 56–150 Minutes of APC 1933–1941, APCR.

75. APC meeting minutes, 19 Jan. 1927, 56–149 Minutes of APC 1927–1933, APCR.

76. APC meeting minutes, 19 June 1935, 56–150 Minutes of APC 1933–1941, APCR. Dawson was contacted on 10 April and filed his report with the Commission on 31 May.

77. APC meeting minutes, 19 June 1935, 56–150 Minutes of APC 1933–1941, APCR.

78. Lavinius L. Williams Esq. to the Audubon Park Commission, 30 Dec. 1933, Manuscript SS 56, V1.4, Box 9, Special Collections, Earl K. Long Library; H.O. Barker to F.E. Neelis, 27 Dec. 1933, Manuscript SS 56, V1.4, Box 9, Special Collections, Earl K. Long Library.

79. "Minutes of Monthly Meeting, APC, 15 Jan. 1936, Audubon Park Collection, 56–150 Minutes of APC 1933–1941. The Bridle Club was rebuilt in a different location; see Audubon Park Commission annual meeting minutes, 20 Jan. 1937, 56–150 Minutes of APC 1933–1941, APCR.

80. APC meeting minutes, 18 Oct. 1935, Dec. 1935, 11 Nov. 1935, 56–150 Minutes of APC 1933–1941, APCR.

81. APC meeting minutes, 15 Jan. 1936, 17 Nov. 1937, 27 June 1935, 56–150 Minutes of APC 1933–1941, APCR.

82. "Merzes Honored at Dedication of Zoo at Audubon," *Times Picayune*, 16 May 1938, New Orleans Public Library; David Niles to Joachim O. Fernandez, 18 Jan. 1939, Mayor Robert Sidney Maestri Records 1936–1946, Box 6, Folder 2 (Works Progress Administration), City Archives, New Orleans Public Library.

83. APC meeting minutes, 19 April 1939, 21 June 1939, 15 Nov. 1939, 56–150 Minutes of APC 1933–41; Superintendent's Reports, 20 Nov. 1940 and 16 April 1941, 56–1 Annual Reports 1940–41, APCR.

84. Robert S. Maestri, no title, *The American*, 25 Sept. 1937, Mayor's Office, Newspaper Clippings from the Administration of Robert S. Maestri, vol. 4, City Archives, New Orleans Public Library; "WPA Administrator Lauds Mayor on First Year of Work," *The*

Sunday Item Tribune, August 1937, Mayor's Office, Newspaper Clippings from the Administration of Robert S. Maestri, vol. 3, City Archives, New Orleans Public Library; "New Homes Are Built For Animals," *Sunday Item-Tribune*, 13 Aug. 1937, New Orleans Public Library.

85. Superintendent's Annual Report [Audubon Park], 19 Jan. 1938, Robert Sidney Maestri Collection 1936–1945, Carton 6, Folder 1 (Works Progress Administration), New Orleans Public Library.

Chapter 2

1. Clarence Cottam to Lucile Q. Mann, 17 Oct. 1960, RU 7293, Box 4, Folder 8, Smithsonian Institution Archives (hereafter SIA).

2. C. W. Coats to Lucile Q. Mann, 14 Oct. 1960, RU 7293, Box 4, Folder 8, SIA.

3. W. M. Storther to William Mann, 9 Dec. 1927, W. M. Box 4, Folder 8, SIA.

4. Pamela Henson. Lucile Quarry Mann Transcript, June–August 1977, Box 1 RU 9513, p. 9, SIA.

5. William M. Mann, *Ant Hill Odyssey*, 4–6.

6. Pamela Henson, Lucile Quarry Mann Transcript, June–August 1977, Box 1 RU 9513, pp. 192–93, SIA; Mann, *Ant Hill Odyssey*, 11–20.

7. Mann, *Ant Hill Odyssey*, 21–22; William Mann to Anna Mann, 1 June 1902 and William Mann to Anna Mann, 10 Oct. 1902, RU 7293, Box 2, SIA.

8. Mann, *Ant Hill Odyssey*, 52.

9. Mann, *Ant Hill Odyssey*, 56–57.

10. William Morton Wheeler, 43–147, 47.

11. Mann, *Ant Hill Odyssey*, 60–61.

12. Mann, *Ant Hill Odyssey*, 79, 81–82.

13. William Mann to Anna Mann, 25 March 1911; William Mann to Anna Mann, 23 July 1911; William Mann to Anna Mann, 20 July 1911, RU 7293, Box 2, SIA.

14. William Mann to Anna Mann, 30 Oct. 1911, RU 7293, Box 2, SIA; William M. Mann, *Ant Hill Odyssey*, 112.

15. Pamela Henson. Lucile Quarry Mann Transcript, June–August 1977, Box 1 RU 9513, p. 10, SIA.

16. Pamela Henson, Lucile Quarry Mann Transcript, June–August 1977, Box 1 RU 9513, p. 174, 168 SIA; Thomas Crosby, *A Zoo for All Seasons: The Smithsonian Animal World*, 35.

17. Pamela Henson, Lucile Quarry Mann Transcript, 4, 5–6, 16, 19–21, SIA; speakeasy detail from Lucile Mann, *Mann in the Zoo, The Houses That Shine*, unpublished manuscript, Box 8, RU 7293, SIA; Pamela Henson, Lucile Quarry Mann Transcript, 222–23, 228, SIA.

18. Pamela Henson, Lucile Quarry Mann Transcript, 14, 21, SIA.

19. Lucile Mann, *From Jungle to Zoo*, 246, 42.

20. Mann, *From Jungle to Zoo,* 246, 42, 26–27, 5.

21. Pamela Henson, Lucile Quarry Mann Transcript, 31–32, SIA.

22. Pamela Henson, Lucile Quarry Mann Transcript, 229–30, SIA.

23. Doris Mable Cochran, Folder 17, Box 7, Doris Mabel Cochran Papers, 1891–1968, RU 7151, SIA; Doris Cochran, Meeting Minutes, 1925–1930, Box 1, Vivarium Society, RU 7163, SIA.

24. Doris Cochran, Meeting Minutes, 14 July 1925, Box 1, Vivarium Society, RU 7163, SIA.

25. Doris Cochran, Meeting Minutes, 17 July and 18 Dec. 1925, Box 1, Vivarium Society, RU 7163, SIA.

26. Doris Cochran, Meeting Minutes, 18 Sept. 1930, Box 1, Vivarium Society, RU 7163, SIA.

27. "Misrepresentation of Jungle Film Denounced by Authorities," Folder 4, Box 251, RU 74, SIA.

28. See RU 74, Box 251, F 3 and F 5, SIA.

29. Pamela Henson, Lucile Quarry Mann Transcript, 61–62, SIA; Vernon Kisling, 165, 168.

30. Henry C. Muskoff (Secretary) to Unknown, no date (1920s), Zoological Society folder, Box 13, Park and Recreation General Subject Files, 1911–1960, Dallas Municipal Archives, Office of the City Secretary, City of Dallas, Texas.

31. George P. Vierheller. "Foreword," in *Zoological Parks and Aquariums*, ed. Will O. Doolittle (American Association of Zoological Parks and Aquariums, 1932), v.

32. Doolittle. "Introduction," in *Zoological Parks and Aquariums*, xi.

33. William T. Hornaday, "Observations on Zoological Park Foundations," in *Zoo-*

logical Parks and Aquariums, 4. Many of the contributions to this volume had previously appeared in *Parks and Recreation* or zoological society magazines between 1924 and 1930. Thus, they do not address the effects of the Depression.

34. Hornaday, 4–5.

35. "Ernest Untermann Passes," *The Vernal Express*, 5 Jan.1956, Ernest Untermann Papers [microform], 1933–1956, Milwaukee Micro Collection 23, University Manuscript Collection, University of Wisconsin-Milwaukee Libraries (hereafter Ernest Untermann Papers, UWM).

36. "How the Zoo Differs From Other Park Departments," 14 June 1937, Ernest Untermann Papers, UWM.

37. Statement made before joint meeting of Civil Service and Park Commission, 25 July 1938, Untermann Papers, UWM.

38. ibid.

39. ibid.

40. ibid.

41. ibid.

42. Hornaday, 5; Statement made before joint meeting of Civil Service and Park Commission, 25 July 1938, Untermann Papers, UWM.

43. Will O. Doolittle. "Introduction," in *Zoological Parks and Aquariums* (American Association of Zoological Parks and Aquariums, 1932), xi.

44. Fairfield Osborn, "The Zoological Society in These Times," *Science* (21 March 1941), 269.

45. Roger Conant, "The Educational Duty of the Zoological Park," in *Zoological Parks and Aquariums*, ed. Doolittle, 15–17.

46. The manuscript of this speech is undated, but the Milwaukee County Zoo suggests the date and place in their history: <http://www.milwaukeezoo.org/students/history/washingtonparkzoo.php>.

47. Ernest Untermann, "Modern Zoo Problems," no date, p. 1, Untermann Papers, UWM.

48. Untermann, "Modern Zoo Problems," 3.

49. ibid, 4.

50. "Central Park Zoo to Get New Houses," *New York Times*, 20 Feb. 1934.

51. *News Bulletin of the Staten Island Zoological Society*; "Zoo to Spur A Hobby Era," *New York Times*, undated clipping, Staten Island Zoo Archives.

52. "Fellowships of the Zoological Society of San Diego," *Science* (19 July 1940), 55; "The Biological Research Institute of the Zoological Society of San Diego," *Science* (6 Sept. 1940), 212.

53. Roger Conant, *A Field Guide to the Life and Times of Roger Conant*, 137–39; Fairfield Osborn, "The Zoological Society in These Times," *Science* (21 March 1941), 269–70.

54. *News Bulletin of the Staten Island Zoological Society* (December 1934), 1.

55. "Lehman Signs Bill for Zoo," *Staten Island Advance*, 7 May 1935; "Governor Presents Beaver to Island Zoo," *Staten Island Advance*, 14 May 1935, Staten Island Zoo scrapbooks, Staten Island Zoo Archives.

56. Alfred Luehrmann, "Value of a Zoological Society," in *Zoological Parks and Aquariums*, ed. Doolittle, 12.

57. *News Bulletin of the Staten Island Zoological Society*, December 1934, 2.

58. "Jungle Ball Parties Listed," *Staten Island Advance*, 29 Nov. 1935; Jungle Ball menu, Staten Island Zoo scrapbooks, Staten Island Zoo Archives; "Animals Greet Social Elite at Jungle Ball on Pier 6," *Staten Island Advance*, 2 Dec. 1935, Staten Island Zoological Society Records, Staten Island Museum.

59. "$30,450 Given for Museum," *Staten Island Advance*, 18 July 1935, Staten Island Zoo scrapbook, Staten Island Zoo Archives.

60. See, for example, "Enough Venom to Kill 140 Men Extracted From Writhing Cobra," *Herald Tribune*, 19 July 1935, Staten Island Zoo scrapbook, Staten Island Zoo Archives. Frank Dickson's name was often misspelled as Dixon.

61. "Stryker Named Head of New Zoo," *Staten Island Advance*, 25 July 1935, Staten Island Zoo scrapbook, Staten Island Zoo Archives.

62. "City Called Duty-Bound to Furnish Zoo Funds," *Staten Island Advance*, 20 July 1936, Staten Island Zoo scrapbooks, Staten Island Zoo Archives.

63. "The Biological Research Institute of the Zoological Society of San Diego," *Science* (6 Sept. 1940), 212.

64. Ken Kawata, *New York's Biggest Little Zoo: A History of the Staten Island Zoo*, 30;

"Stryker Named Head of Zoo," *Staten Island Advance*, 25 July 1936.

65. "Zoo Asks for Increase in Budget," *Staten Island Advance*, 2 Oct. 1936.

66. "Society Aided by WPA," *News Bulletin of the Staten Island Zoological Society* (Jan. 1937), 4; "Successful First Year Concludes," *News Bulletin of the Staten Island Zoological Society* (July 1937), 4; "WPA Workers Supplement Zoo Staff," *News Bulletin of the Staten Island Zoological Society* (Sept. 1938), 3; "WPA and NYA Active in Zoo," *News Bulletin of the Staten Island Zoological Society* (May 1938), 3–4; "Annual Membership Meeting Held," *News Bulletin of the Staten Island Zoological Society* (March 1939), 3–4; "Society's Annual Report Given," *News Bulletin of the Staten Island Zoological Society* (March 1940), 2, 4.

67. Kawata, *New York's Biggest Little Zoo*, 31.

68. See WPA Federal Writers Project, Record Series 18, 19, 20, and 22, Microfilm rolls 86–89, New York City Municipal Archives.

69. See Gus Knudson's monthly reports from 1933–1936 in Woodland Park Zoo, Monthly and Annual Reports, 1933–1936, 51/6, Seattle Municipal Archives. Quotation is from the report of 28 Feb. 1934.

70. Hornaday, in *Zoological Parks and Aquariums*, 2–3.

71. Doolittle, "Introduction," in *Zoological Parks and Aquariums*, xi.

72. "For Your Information: History of the Woodland Park Zoological Gardens," pp. 2 and 5, no date [probably late 1930s], folder 50/1, Woodland Park and Zoo 1906–1949, Don Sherwood Parks History Collection, Seattle Municipal Archives.

73. Roger Conant, *A Field Guide to the Life and Times of Roger Conant*, 496.

74. ibid, 21–22.

75. *Tulsa Zoo: The First Fifty Years*, 2–4.

76. ibid, 7–8.

77. ibid, 12.

78. "Holzworth, John Michael," reprint from *Who's Who in America*, a clipping taped to the front of Holzworth's book *The Twin Grizzlies of Admiralty Island.*

79. William T. Hornaday. "Forward," in *The Wild Grizzlies of Alaska*, by John M. Holzworth, v.

80. ibid, v.

81. ibid, vi.

82. ibid, vi.

83. John Holzworth, radio address, no date, included in Hearing before the Senate Special Committee on Conservation of Wild Life Resources, *Protection and Preservation of the Brown and Grizzly Bears of Alaska*, 18 Jan. 1932, 17.

84. Ernest Untermann to Dr. Frank Thone, *Science News*, 5 Aug. 1939, p. 90, Untermann Papers, UWM.

85. Untermann, "Modern Zoo Problems."

86. Edward J. Larson.

87. Belle J. Benchley, *My Friends the Apes*, ix.

88. ibid, ix.

89. ibid, x.

90. ibid, xii, emphasis in original.

91. ibid, xii,

92. ibid, 217.

93. Ernest Untermann, "Our Friends at the Zoo," vol. IV, no. 1, January 1940, Ernest Untermann Papers, UWM.

94. ibid.

95. Ernest Untermann, "The Zoo, Old and New," no date, p. 1, Ernest Untermann Papers, UWM.

96. Ernest Untermann, "Our Friends at the Zoo," vol. IV. no. 1, January 1940, Ernest Untermann Papers, UWM.

97. Mr. Fay's Interview of Ernest Untermann, no date, p. 1, Ernest Untermann Papers, UWM.

98. Belle J. Benchley, *My Friends the Apes*, viii.

99. ibid, viii.

100. ibid, viii– ix, emphasis in original.

101. Ernest Untermann, "Our Friends at the Zoo," vol. IV, no. 1, January 1940, Ernest Untermann Papers, UWM.

102. Robert Moses to F. Trubee Davison, 7 Oct. 1940 and 22 Sept. 1940, NYC Dept. of Parks Office of the Commissioner (Robert Moses) 1940–1956, Microfilm Reel 5, Folder 032 — Museum of Natural History, New York City Municipal Archives.

103. Gus Knudson, "Monthly Report," 31 May 1933, p. 1, folder 51/6, Woodland Park Zoo Monthly and Annual Reports 1933–1936, Don Sherwood Parks History Collection, Seattle Municipal Archives.

104. Gus Knudson, "Monthly Report," 30 June 1933, p. 1, folder 51/6, Woodland Park Zoo Monthly and Annual Reports 1933–

1936, Don Sherwood Parks History Collection, Seattle Municipal Archives.

105. Pamela Henson, Lucile Quarry Mann Transcript, 26, 27, SIA.

106. Roger Conant, *A Field Guide to the Life and Times of Roger Conant*, 494.

107. Pamela Henson, Lucile Quarry Mann Transcript, 26, 54–55, SIA.

108. Pamela Henson, Lucile Quarry Mann Transcript, 95–97, 239, SIA; Henry Ringling North to William Mann, 9 Feb. 1940, Box 253, Folder 3, RU 74, SIA; L. C. Holland to Mann, 25 March 1940, Box 253, folder 3, RU 74, SIA. The poem was titled "Circus Day."

109. William Mann to Bill Morden, 7 April 1932, RU 74, Box 251, F 5, SIA;

Pamela Henson, Lucile Quarry Mann Transcript, 95–97, SIA.

Chapter 3

1. Quoted in *Tulsa Zoo: The First Fifty Years* , 11.

2. "Moses Rebukes a Critic," *New York Times*, 26 Feb. 1976, ProQuest Historical Newspapers.

3. "Two Projects Outrage Doctor's Sense of Proportion," *Staten Island Advance*, 29 Dec. 1933, Staten Island Zoological Society Records, Staten Island Museum.

4. "New Bear Pits, Monkey Island Entrance Throngs at S. Antonio," no source, 4 Nov. 1929, St. Louis Zoo Records, 1910–1940, Roll 6, Western Historical Manuscript Collection, University of St. Louis-Missouri.

5. Detroit Zoological Society, *Wonders Among Us: Celebrating 75 Years at the Detroit Zoo* (Detroit: Detroit Zoological Society, 1974), 26.

6. William A. Austin, 26

7. "Zoo Visitors Here Setting Records," *New York Times*, 17 Nov. 1935, ProQuest Historical Newspapers.

8. "$2,000,000 Zoo Cost Denied by Moses," *New York Times*, 17 April 1935, ProQuest Historical Newspapers.

9. Robert Moses, *Public Works: A Dangerous Trade*, 702.

10. See, for example, "New Homes Are Built For Animals," *Sunday Item-Tribune*, 13 Aug. 1937, New Orleans Public Library.

11. For a history of the American welfare state see Michael Katz, *In the Shadow of the Poorhouse: A Social History of Welfare in America*.

12. "Picture-Book Zoo Being Built in Park," *New York Times*, 9 March 1934, ProQuest Historical Newspapers.

13. "Park Work Relief to Speed Projects," *New York Times*, 31 Dec. 1934, ProQuest Historical Newspapers.

14. "$80,000,000 Spent by CWA in This State; 240,000 Workers Off Federal Payroll Today," *New York Times*, 2 April 1934, ProQuest Historical Newspapers.

15. "Park Exposition Opens Tomorrow," *New York Times*, 9 June 1935, ProQuest Historical Newspapers.

16. L. H. Robbins, "Our Park Creator Extraordinary," *New York Times*, 20 March 1938, ProQuest Historical Newspapers.

17. "Expositions," *News Bulletin of the Staten Island Zoological Society* (Nov. 1936), 2.

18. "For Your Information: History of the Woodland Park Zoological Gardens," p. 2, no date [probably late 1930s], folder 50/1, Woodland Park and Zoo 1906–1949, Don Sherwood Parks History Collection, Seattle Municipal Archives.

19. Newspaper clipping, no source or date, St. Louis Zoo Records, 1910–1940, Roll 6, Western Historical Manuscript Collection, University of St. Louis-Missouri (hereafter WHMC).

20. Such a campaign failed in New Orleans, for example.

21. Newspaper clipping, no source or date, St. Louis Zoo Records, 1910–1940, Roll 6, WHMC.

22. Technically, the city "took" the special tax by not levying it and instead upping the general levy eight cents to the maximum $1.35 per $106 of property value.

23. "Suit to Block Move to Divert Special Taxes," no date or source, St. Louis Zoo Records, 1910–1940, Roll 6, WHMC.

24. Title and author illegible, *St. Louis Star*, 10 Aug. 1927, St. Louis Zoo Records, 1910–1940, Roll 6, WHMC.

25. "Zoo, with $235,000 Assured Yearly, Now Safe, Dieckman Says," *The Globe-Democrat*, 19 Jan. 1928, St. Louis Zoo Records, 1910–1940, Roll 6, WHMC.

26. Title and author illegible, *St. Louis Star*, 10 Aug. 1927, St. Louis Zoo Records, 1910–1940, Roll 6, WHMC.

27. Title and author illegible, *St. Louis*

Star, 10 Aug. 1927, St. Louis Zoo Records, 1910–1940, Roll 6, WHMC.

28. No title or author, *St. Louis Star*, 1927, no exact date, St. Louis Zoo Records, 1910–1940, Roll 6, WHMC.

29. "Suit to Block Move to Divert Special Taxes," no date or source, St. Louis Zoo Records, 1910–1940, Roll 6, WHMC.

30. "Zoological Society to Sue to Retain Tax," no source, 1 May 1927, St. Louis Zoo Records, 1910–1940, Roll 6, WHMC.

31. "Zoological Board Will Fight Special Tax Abolishment," no source or date, St. Louis Zoo Records, 1910–1940, Roll 6, WHMC.

32. "City Ordered to Set Aside Library and Zoo Funds from Tax," *The Globe-Democrat*, 10 Jan. 1928, St. Louis Zoo Records, 1910–1940, Roll 6, WHMC.

33. "The Zoo Tax is Upheld," *The St. Louis Star*, no date, St. Louis Zoo Records, 1910–1940, Roll 6, WHMC.

34. No title or author, *The St. Louis Star*, no date, St. Louis Zoo Records, 1910–1940, Roll 6, WHMC.

35. "The City's Revenue Plight," *The St. Louis Star*, June 1927, St. Louis Zoo Records, 1910–1940, Roll 6, WHMC.

36. Title and source illegible, 1932, St. Louis Zoo Records, 1910–1940, Roll 6, WHMC.

37. "Zoo is Nailed Down," *St. Louis Star*, March 1932, no author, St. Louis Zoo Records, 1910–1940, Roll 6, WHMC.

38. Untitled editorial from *The Globe-Democrat*, 7 Aug. 1931, St. Louis Zoo Records, 1910–1940, Roll 6, WHMC.

39. "City Hall...Fight on Special Taxes for Art Museum and Zoo," *Post*, 16 Sept. 1932, St. Louis Zoo Records, 1910–1940, Roll 6, WHMC.

40. ibid.

41. Untitled editorial from *The Globe-Democrat*, 7 Aug. 1931, St. Louis Zoo Records, 1910–1940, Roll 6, WHMC.

42. Untitled editorial, *St. Louis Star*, 29 Aug. 1932, St. Louis Zoo Records, 1910–1940, Roll 6, WHMC.

43. Erwin O. Schneider, no title, *St. Louis Star and Times*, no date, St. Louis Zoo Records, 1910–1940, Roll 6, WHMC.

44. "Normal Domestic Tourist Travel in 1930 is Forecast," *Washington Post*, 11 April 1930, St. Louis Zoo Records, 1910–1940, Roll 6, WHMC.

45. George Vierheller, "Zoo Spreads Fame of St. Louis Around the Globe, Director Says," *St. Louis Star*, no date, St. Louis Zoo Records, 1910–1940, Roll 6, WHMC.

46. "For Your Information: History of the Woodland Park Zoological Gardens," p. 1, no date [probably late 1930s], folder 50/1, Woodland Park and Zoo 1906–1949, Don Sherwood Parks History Collection, Seattle Municipal Archives (hereafter DSPHC).

47. See Knudson's monthly reports (averaging 4 pages each) in folder 51/6, Woodland Park Zoo Annual Reports 1933–1936, DSPHC.

48. Superintendent's Report, 20 March 1940, Audubon Park Commission Records, 56-, Annual Reports 1940–41, Earl K. Long Library, University of New Orleans.

49. "Mayor Thinks Move Would Get More Tourists," no source or date, St. Louis Zoo Records, 1910–1940, Roll 6, WHMC.

50. Untitled Editorial, *St. Louis Star*, 29 Aug. 1932, St. Louis Zoo Records, 1910–1940, Roll 6, WHMC.

51. Erwin O. Schneider, untitled and undated editorial, *St. Louis Star and Times*, St. Louis Zoo Records, 1910–1940, Roll 6, WHMC.

52. "The Zoo and the Art Museum," *Post Democrat*, 27 Aug. 1932, St. Louis Zoo Records, 1910–1940, Roll 6, WHMC.

53. "Zoo is Nailed Down," no author, *Times*, January 1932, St. Louis Zoo Records, 1910–1940, Roll 6, WHMC.

54. "The Zoo and the Art Museum," *Post Democrat*, 27 Aug. 1932, St. Louis Zoo Records, 1910–1940, Roll 6, WHMC.

55. "Just Another Way Around," 26 Feb. 1932, *The Globe Democrat*, St. Louis Zoo Records, 1910–1940, Roll 6, WHMC.

56. "Zoo is Nailed Down," *St. Louis Star*, March 1932, no author, St. Louis Zoo Records, 1910–1940, Roll 6, WHMC.

57. Ken Kawata, personal communication, Feb. 2007. The director was Hugh Davis (Oklahoma).

58. "Bear on Rampage Routs 100 at Zoo," *New York Times*, 3 May 1934, ProQuest Historical Newspapers.

59. Robert Moses, *Public Works: A Dangerous Trade*, 686.

60. *The Emergency Work Relief Program of the F.E.R.A., April 1, 1934—July 1, 1935*, 30. Emphasis added.

61. *Annual Report of the Department of*

Parks, Borough of Manhattan, 1931, p. 24, New York City Department of Parks and Recreation Library.

62. Robert Moses to William N. Matthews, 18 Jan. 1935, box 11, folder E, Manuscripts and Archives Section, New York Public Library (hereafter NYPL).

63. Public Hearing Before Aldermanic Committee Investigating Relief, 8 May 1935, p. 1982, box 11, folder P, NYPL.

64. Public Hearing Before Aldermanic Committee Investigating Relief, 8 May 1935, p. 1986, box 11, folder P, NYPL.

65. Public Hearing Before Aldermanic Committee Investigating Relief, 8 May 1935, p. 1988, box 11, folder P, Robert Moses Papers, NYPL.

66. Robert Moses, memo to mayor, 10 Sept. 1935, Box 97, NY-Parks Dept 1935, Robert Moses Papers, NYPL.

67. Public Hearing Before Aldermanic Committee Investigating Relief, 8 May 1935, p. 1989, box 11, folder P, Robert Moses Papers, NYPL.

68. "Smith Now 'Agent' for Prospect Zoo," *New York Times*, 10 May 1935, ProQuest Historical Newspapers.

69. Robert Caro, 368, 370 (quotation), 372–73.

70. "Al Smith Named Curator of New Zoo," *Staten Island Advance*, 11 June 1936, Staten Island Zoo scrapbooks, Staten Island Zoo Archives.

71. Gus Knudson, "Monthly Report," 30 April 1933, p. 5, folder 51/6, Woodland Park Zoo Monthly and Annual Reports 1933–1936, DSPHC, SMA.

72. See, for example, Ed Bean to Stanley Field, 3 Jan. 1934, Grounds and Building Committee 1934, Box 90, Brookfield Zoo Archives.

73. "Bricklayers Begin Work on Park Zoo," *New York Times*, no date, Central Park Zoo file, New York City Department of Parks and Recreation Library.

74. "Exposition Opens With Welcome by Palma," *Staten Island Advance*, 18 Oct. 1935, Staten Island Zoo scrapbook, Staten Island Zoo Archives; "15,000 Jam Exposition at Pier Six," *Staten Island Advance*, 22 Oct. 1935, Staten Island Zoo scrapbook, Staten Island Zoo Archives.

75. "'Picture Book' Zoo Will Open Today," newspaper clipping, 1935, Lillian Swann Saarinen Papers 1924–1974, Microfilm Roll 1152, Archives of American Art.

76. "New Zoo Opens," *New York Times*, 3 Dec. 1934, ProQuest Historical Newspapers.

77. "'Pep' Talk by Smith Opens Zoo Drive," *New York Times*, 28 May 1935, ProQuest Historical Newspapers.

78. "Smith Decries 'Back-Alley Politics' of La Guardia in Row With Moses," *New York Times*, 4 July 1935, ProQuest Historical Newspapers.

79. "From Our Readers: Zoo Project Brings Echoes of General Satisfaction," *Staten Island Advance*, 13 March 1934, Staten Island Zoo scrapbooks, Staten Island Zoo Archives.

80. Howard Braucher to Members of the American Institute of Park Executives and the American Park Society, March 1932, folder 4/16, Recreation and Education 1932, DSPHC.

81. Robert D. Leighninger Jr., 30.

82. *The Emergency Work Relief Program of the F.E.R.A., April 1, 1934—July 1, 1935*, 88.

83. Harry Hopkins, *Inventory: An Appraisal of Results of the Works Progress Administration*, 19–20.

84. Howard Braucher to Members of the American Institute of Park Executives and the American Park Society, March 1932, folder 4/16, Recreation and Education 1932, DSPHC.

85. Harry Hopkins, *Spending to Save: The Complete Story of Relief*, 122, 177.

86. Phoebe Cutler, 8–9.

87. Howard Braucher to Members of the American Institute of Park Executives and the American Park Society, March 1932, folder 4/16, Recreation and Education 1932, DSPHC.

88. Robert Moses, "You Can Trust the Public," *The American Magazine* (July 1939): 144–45, 58–59. Quotation from p. 145.

89. "Moses Defends High Park Costs," *New York Times*, 20 Jan. 1940, ProQuest Historical Newspapers.

90. "Seattle Receives Pet," newspaper clipping, 1922, folder 16/5, Woodland Park and Zoo 1922–1939, Ben Evans Recreation Program History Collection, Seattle Municipal Archives; "For Your Information: History of Woodland Park Zoological Gardens," no date [probably late 1930s], folder 50/1, Woodland Park and Zoo 1906–1949, DSPHC.

91. "New Bear Pits, Monkey Island Entrance Throngs at S. Antonio," no source, 4 Nov. 1929, St. Louis Zoo Records, 1910–1940, Roll 6, WHMC.

92. Wilbur L. Matthews, *History of the San Antonio Zoo*, 15.

93. "Zoos Popular Fun is Cheap: Attendance at Gardens Showing New Records in Backward Times," no date or source, St. Louis Zoo Records, 1910–1940, Roll 6, WHMC.

94. "Zoo Attendance a Record," *New York Times*, 24 April 1935, ProQuest Historical Newspapers; Milton Bracker, "From Aoudad to Zebra," *New York Times*, 26 Nov. 1939, ProQuest Historical Newspapers. Other big-city zoos of the era claimed similar attendance figures, although zoo historians are somewhat skeptical about their accuracy. The exact number of visitors seems less important, however, than the contemporary *perception* that zoos served so many citizens.

95. "Zoo Visitors Here Setting Records," *New York Times*, 17 Nov. 1935, ProQuest Historical Newspapers.

96. "Annual Membership Meeting Held," *News Bulletin of the Staten Island Zoological Society* (March 1939): 3.

97. Frank Neelis report, 6 Sept. 1939, Audubon Park Commission Records, Box 154 Reports of Zoological Society, Earl K. Long Library, University of New Orleans.

98. "For Your Information: History of the Woodland Park Zoological Gardens," no date [probably late 1930s], folder 50/1, Woodland Park and Zoo 1906–1949, DSPHC.

99. *The Emergency Work Relief Program of the F.E.R.A., April 1, 1934—July 1, 1935*, 95.

100. Gus Knudson, "Monthly Report," 31 May 1935, p. 5, folder 51/6, Woodland Park Zoo Monthly and Annual Reports 1933–1936, DSPHC.

101. "Smith Decries 'Back-Alley Politics' of La Guardia in Row With Moses," *New York Times*, 4 July 1935, ProQuest Historical Newspapers.

102. Ellis Joseph to J. W. Thompson, 17 March 1920, folder 50/5, Woodland Park Zoo Animals — Popular/Totem 1916–1949, DSPHC.

103. Jonathan Kuhn to Gary Zarr, Memorandum on Central Park Zoo History and Evolution, p. 3, Zoos file, New York City Department of Parks and Recreation Library; *Annual Report, Department of Parks, City of New York* (1904), p. 37, New York City Department of Parks and Recreation Library.

104. "Audubon Park" [press release], 26 May 1937, Audubon Park Commission Records, 56–142 Minutes of Audubon Park Commission, Earl K. Long Library, University of New Orleans.

105. Unknown to Mayor Sawnie R. Aldredge, 13 Jan. 1922, Mayor file, Box 6, Park and Recreation Department General Subject Files, 1911–1970, Dallas Municipal Archives, Office of the City Secretary, City of Dallas, Texas.

106. W. F. Jacoby to Park Board, 2 Jan. 1930, Zoo Report folder, Box 13, Park and Recreation Department General Subject Files, 1911–1970, Dallas Municipal Archives, Office of the City Secretary, City of Dallas, Texas.

107. Paul Masserman, "Zoo Gains Favor Rapidly as Recreation," *The Evansville Courier and Journal*, 27 April 1930, St. Louis Zoo Records, 1910–1940, Roll 6, WHMC.

108. Gus Knudson, "Monthly Report," 30 Nov. 1933, p. 1, folder 51/6, Woodland Park Zoo Monthly and Annual Reports 1933–1936, DSPHC.

109. Gus Knudson, "Monthly Report," 31 March 1934, p. 2, folder 51/6, Woodland Park Zoo Monthly and Annual Reports 1933–1936, DSPHC.

110. Gus Knudson, "Monthly Report," 28 Feb. 1934, p. 1, folder 51/6, Woodland Park Zoo Monthly and Annual Reports 1933–1936, DSPHC.

111. "Staten Island Zoo to Open Wednesday," *New York Times*, 7 June 1936, ProQuest Historical Newspapers; "Staten Island Zoo to be Opened Today," *New York Times*, 10 June 1936, ProQuest Historical Newspapers.

112. "Smith Opens A Zoo and Gets a Third Job," *New York Times*, 11 June 1936, ProQuest Historical Newspapers; "Al Smith Named Curator of New Zoo," *Staten Island Advance*, 11 June 1936, Staten Island Zoo scrapbooks, Staten Island Zoo Archives.

113. "Frogs Hop Lazily in Jubilee Test," *New York Times*, 19 May 1935, ProQuest Historical Newspapers.

114. "Smith Gives Names to 46 Zoo Animals," *New York Times*, 28 Sept. 1935, ProQuest Historical Newspapers.

115. Data taken from L.C. Everard.

116. Andrea Ross, 41, 45, 50.

117. *The Emergency Work Relief Program of the F.E.R.A., April 1, 1934—July 1, 1935*, 103, 111.

118. Harry L. Hopkins, *Report on Progress of the Works Program, October 15, 1936*, 26.

For a description of the New Deal Projects at the Toledo Zoo see Ted Ligibel, 46–83.

119. Robert Moses, "Mr. Moses Surveys the City's Statues," *New York Times*, 21 Nov. 1943, ProQuest Historical Newspapers.

120. Robert Bean to Stanley Field, 21 March 1939, Grounds and Building Committee 1939, Box 91, Brookfield Zoo Archives.

121. Figure comes from Richard C. Berner, 214; Gus Knudson, "Monthly Report," 31 Aug. 1933, pp. 1–2, folder 51/6, Woodland Park Zoo Monthly and Annual Reports 1933–1936, DSPHC, SMA.

122. Gus Knudson, "Monthly Report," 31 July 1934, p. 3, folder 51/6, Woodland Park Zoo Monthly and Annual Reports 1933–1936, DSPHC.

123. Gus Knudson, "Monthly Report," 30 April 1934, p. 1, folder 51/6, Woodland Park Zoo Monthly and Annual Reports 1933–1936, DSPHC.

124. For documentation on Dallas's segregated system, see the following folders: "Exall Park (use by Negro Children), 1950," Box 2; "Golf Courses for Negroes, 1946–1947," Box 3; "Moore Negro Park (8th Street Park) Original Purchase and Expansion, 1937–1945" and "Moore Park 1938–1942 (golf course for Negroes)," Box 6; all in Park and Recreation General Subject Files, 1911–1960, Dallas Municipal Archives, Office of the City Secretary, City of Dallas, Texas.

125. L.B. Houston to Lynne W. Landrum, 23 June 1941, General D folder, Box 2, Park and Recreation Department General Subject Files, 1911–1970, Dallas Municipal Archives, Office of the City Secretary, City of Dallas, Texas.

126. Rich Buickerood, 9.

127. Buickerood, 7; "Zoo Commission Meeting," 24 Jan. 1930, Zoological Society folder, Box 13, Park and Recreation Department General Subject Files, 1911–1970, Dallas Municipal Archives, Office of the City Secretary, City of Dallas, Texas.

128. Edward Richard O'Neill to Mayor Robert Maestri, 28 April 1941, Audubon Park Commission Records, Box 56–9 Correspondence, Earl K. Long Library, University of New Orleans (hereafter EKLL).

129. "Audubon Park" [press release], 26 May 1937, Audubon Park Commission Records, 56–142 Minutes of Audubon Park Commission, EKLL.

130. Audubon Park Commission minutes, 15 Feb. 1939, 56–149, Minutes of Audubon Park Commission 1933–1941, EKLL.

131. "Audubon Park" [press release], 26 May 1937, Audubon Park Commission Records, 56–142 Minutes of Audubon Park Commission, EKLL.

132. Local Women's Business Organizations to Mayor Maestri, 56–9, Correspondence, January 1940, Louisiana Special Collections, EKLL.

133. Anonymous Letter to E. Neelis, 20 April 1940, Box 9, Correspondence, Louisiana Special Collections, EKLL.

134. Louis A.G. Blanchet to Mayor Maestri, 26 July 1940, Box 9, Correspondence, Louisiana Special Collections, EKLL.

135. Nancy Weiss, 35.

136. ibid, 53.

137. Bonnie Fox Schwartz, 216–17, 229.

138. Weiss, 53.

139. "Minutes of Monthly Meeting, Audubon Park Commission, 15 Jan. 1936, Audubon Park Collection, 56, F 150, Minutes, APC 1933–1941, EKLL.

140. U.S. Works Progress Administration, Civil Works Projects, Labor and Building Costs, 1933, Audubon Park Commission Records, 56–63, EKLL.

141. Audubon Park Commission Annual Meeting Minutes, 15 Jan. 1936, 56–150 Minutes of APC 1933–1941, Audubon Park Commission Collection, EKLL.

142. Minutes of Audubon Park Commission, 15 Jan. 1941 and 19 Feb. 1941, Audubon Park Commission Records, 56–142, Minutes of APC 1940–41, EKLL.

143. Minutes of Audubon Park Commission, 21 May 1941, Audubon Park Commission Records, 56–142, Minutes of APC 1940–41, EKLL.

144. George Lewis and Byron Fish, 38–39.

145. J.L. Buck to Floyd S. Young, 30 Aug. 1939, Marlin Perkins Papers, 516, F49, Box 3, Correspondence Regarding Lincoln Park Zoo, 1939–1962, WHMC. The African woman is described as a wet nurse in Mark Rosenthal, et al., 4.

146. ibid.

147. "Ellis Stanley Joseph" [obituary], *Science*, 21 Oct. 1938, 368–69.

148. Marlin Perkins Papers, 516, F 69, Box 4, Postcards, 1930–1978, WHMC.

149. Harold C. Bingham, "Sex Development in Apes," 5. Quoted from H.C. Chapman, "On the Structure of the Chimpanzee," Proc Acad. Nat. Sci., Philadelphia, 1879, 52–63, 58.

150. "Monkeyshines," no source, 21 Aug. 1940, folder 16/6, Woodland Park and Zoo 1940–1951, Ben Evans Recreation Program History Collection, Seattle Municipal Archives.

151. "Chief Seattle Proves They Can Come Back," no date or source, Gustave Knudson clippings 1910s–1930s, Woodland Park Zoo Archives.

152. E. R. O'Neill to A. Labas, 22 May 1940, Audubon Park Commission Records, Subseries V. 2, Box 9 Correspondence, EKLL.

153. Andrea Ross, 47.

154. *Report of the Board of Park Commissioners of the City of Seattle 1923–1930*, pp. 100–101, folder 4/4, Annual Report 1923–30, DSPHC.

155. "For Your Information: History of the Woodland Park Zoological Gardens," p. 2, no date [probably late 1930s], folder 50/1, Woodland Park and Zoo 1906–1949, DSPHC.

156. "Recreation Survey," 1937, folder 5/12, Correspondence and Financial Records (1937), DSPHC.

157. For a complaint about profanity, see George Hill to Honorable Board of School Directors, 21 Oct. 1919, folder 50/1, Woodland Park and Zoo 1906–1949, DSPHC.

158. "Boy Appointed Park Sheriff," no source, 1921, Clippings from Knudson's Scrapbook, 1910s–1930s, Woodland Park Zoo Archives.

159. *Annual Report 1945, Board of Park Commissioners*, Seattle Park Department Annual Reports 1936–1950 (Excerpts Relating to Woodland Park Zoo), Woodland Park Zoo Archives.

160. Robert Moses to Laurence Rockefeller, 16 Aug. 1940, NYC Dept. of Parks Office of the Commissioner (Robert Moses) 1940–1956, Microfilm Reel 8, Folder 069 — Zoological Society (New York), New York City Municipal Archives.

161. "Boy Appointed Park Sheriff," Newspaper clipping, 1921, no source, clippings from Knudson's Scrapbook, 1910s–1930s, Woodland Park Zoo Archives.

162. Gus Knudson, "Monthly Report," 31 Jan. 1935, p. 4, folder 51/6, Woodland Park Zoo Monthly and Annual Reports 1933–1936, DSPHC.

163. Gus Knudson, typescript report of zoo tour, June 1930, p. 25, folder 51/8, Woodland Park Zoo Reports on Tours of Other 1940–1948, DSPHC, SMA.

164. Gus Knudson, "Monthly Report," 31 Oct. 1933, p. 4, folder 51/6, Woodland Park Zoo Monthly and Annual Reports 1933–1936, DSPHC, SMA.

165. E. R. Hoffman to R. H. Case, 31 July 1931, and Simon Burnett (Park Board) to Louis J. Forbes (Chief of Police), 21 April 1931, folder 4/9, Correspondence and Financial Records 1931, DSPHC, SMA.

166. Samuel Martin (President of Board of Park Commissioners) to Chief William H. Sears, 20 Nov. 1936, folder 5/10, Correspondence and Financial Records, DSPHC, SMA.

167. Gus Knudson, typescript report of zoo tour, June 1930, p. 23, folder 51/8, Woodland Park Zoo Reports on Tours of Other 1940–1948, DSPHC, SMA; littering detail from untitled 1937 memo on Seattle's park system, folder 5/12, Correspondence and Financial Records (1937), DSPHC.

168. Allyn Jennings, untitled memo, 26 April 1939, Office of General Director, Allyn Jennings, Wildlife Conservation Society Archives, New York.

169. Gus Knudson, typescript report of zoo tour, June 1930, p. 23, folder 51/8, Woodland Park Zoo Reports on Tours of Other 1940–1948, DSPHC, SMA.

170. "Zoo Facts," *In Animaland* (Nov. 1946): 4.

171. "Baby Puma Saved from Its Parents," *New York Times*, 29 April 1929, ProQuest Historical Newspapers.

172. "For the Protection of Birds and Animals in the Audubon Zoo," no. 7614 Commission Council Series, 28 Nov. 1923, Audubon Park Commission Records, Box 56–8 Audubon Park Commission 1923–1927, EKLL.

173. "Bear Bites Boy, 7, in Central Park," *New York Times*, 27 May 1937, ProQuest Historical Newspapers.

174. Belle J. Benchley, *My Friends the Apes*, 116–117.

175. "Elephant Resents Tail Pulling; Youth Put in Central Park Cell," *New York Times*, 9 April 1932, ProQuest Historical Newspapers.

176. Pamela Henson, Lucile Quarry Mann Transcript, June–August 1977, Box 1 RU 9513, p. 165, SIA.

177. Buickerood, 9.

178. Audubon Park Commission minutes, 18 Oct. 1944, Audubon Park Commission Records, 56–142 Minutes of Audubon Park Commission, EKLL.

179. Annual Report Seattle Park Department 1936, Seattle Park Department Annual Reports 1936–1950 (Excerpts Relating to Woodland Park Zoo), Woodland Park Zoo Archives.

180. "$2500 Prize Zoo Orang-Outang Poisoned by Spectator as He Goes Through Show for Camera," *Globe-Democrat*, 2 July 1935, St. Louis Zoo Records, 1910–1940, Roll 6, WHMC.

181. "Poisoners at the Zoo," *St. Louis Star*, 3 July 1935, St. Louis Zoo Records, 1910–1940, Roll 6, WHMC.

182. Mark Rosenthal, et al., 59.

183. Dogs killed 18 rabbits, four sheep, two goats, and one black swan in one month at the Audubon Zoo, which was not fenced until 1951. Minutes of Audubon Park Commission, 20 Dec. 1944, Audubon Park Commission Records, 56–142, Minutes of APC 1944, EKLL.

184. Gus Knudson, typescript report of zoo tour, June 1930, p. 24, folder 51/8, Woodland Park Zoo Reports on Tours of Other 1940–1948, DSPHC.

185. Gus Knudson, "Zoo Report for 1933," folder 51/6, Woodland Park Zoo Monthly and Annual Reports 1933–1936, DSPHC, SMA.

186. Gus Knudson, "1934 Zoo Report," p. 3, folder 51/6, Woodland Park Zoo Monthly and Annual Reports 1933–1936, DSPHC, SMA.

187. Allyn Jennings to Alfred Ely, 24 May 1940, Office of General Director, Allyn Jennings, Wildlife Conservation Society Archives, New York.

188. Chauncey J. Gordon to Robert Moses, 17 July 1943, NYC Dept. of Parks Office of the Commissioner (Robert Moses) 1940–1956, Microfilm Reel 18, Folder 036 — Zoological Society (New York), New York City Municipal Archives. Moses replied that budget realities precluded Gordon's suggestions; quite simply, even those citizens who demanded free recreational facilities also demanded "cuts in the budget."

189. Pamela Henson, Lucile Quarry Mann Transcript, June–August 1977, Box 1 RU 9513, pp. 176–80, SIA.

190. Marcia Davenport to Robert Moses, 7 March 1944, and Robert Moses to Fairfield Osborn, 8 March 1943, NYC Dept. of Parks Office of the Commissioner (Robert Moses) 1940–1956, Microfilm Reel 18, Folder 038 — Zoological Society (New York), New York City Municipal Archives.

Chapter 4

1. Gus Knudson, "Monthly Report," 30 Sept. 1934, p. 3, folder 51/6, Woodland Park Zoo Monthly and Annual Reports 1933–1936, Don Sherwood Parks History Collection, Seattle Municipal Archives (hereafter DSPHC).

2. William Gaines, *Daily News Miner* (Fairbanks Alaska), 26 Sept. 1931, <http://www.Newspaperarchvies.com> (14 April 2009); *The Ogden Examiner*, Sunday, 1 Jan. 1933, <http://www.newspaperarchives.com> (14 April 2009).

3. Mrs. Charles J. Dolan, "To the Editor of the Post-Dispatch," no date, St. Louis Zoo Records, 1910–1940, Roll 6, Western Historical Manuscript Collection, University of St. Louis-Missouri Archives (hereafter WHMC); "How the Big Zoo Python Eats Every Six Weeks," *San Antonio Light*, 28 Jan. 1934, <http://www.newspaperarchive.com> (9 June 2009).

4. "How the Big Zoo Python Eats Every Six Weeks," *San Antonio Light*, 28 Jan. 1934, <http://www.newspaperarchive.com>, (9 June 2009); "Smith Opens a Zoo and Gets a Third Job," *New York Times*, 11 June 1936, ProQuest Historical Newspapers; "Bring on Those Lions," *Staten Island Advance*, 13 June 1936, Staten Island Zoo scrapbooks, Staten Island Zoo Archives.

5. Newspaper article by Gudrun Andersen, no source or date, Knudson's Scrapbook, Woodland Park Zoo Archives; Gus Knudson, "Monthly Report," 31 May 1933, p. 4, folder 51/6, Woodland Park Zoo Monthly and Annual Reports 1933–1936, DSPHC; Gus Knudson, "Monthly Report," 30 Sept. 1934, p. 3, folder 51/6, Woodland Park Zoo Monthly and Annual Reports 1933–1936, DSPHC; Gus Knudson, "Monthly Report," 31 Aug. 1933, p. 1, folder 51/6, Woodland

Park Zoo Monthly and Annual Reports 1933–1936, DSPHC, SMA; Gus Knudson, "Monthly Report," 30 Sept. 1934, p. 3, folder 51/6, Woodland Park Zoo Monthly and Annual Reports 1933–1936, DSPHC.

6. "Deadly Snakes to Writhe in National Zoo," *Charles City Daily Press* (Iowa), Monday, 22 Dec. 1930; "Reptiles Get Ray Cure," *The Salt Lake Tribune*, Sunday Morning, 26 1931; "All the Comforts of Home for Washington Snakes," *Rock Valley Bee* (Iowa), 5 June 1931; Frederic Haskin, "What do You Want to Know?," *Chester Times* (Chester, Pennsylvania), Monday, 27 March 1937. All newspaper articles in <http://www.newspaperarchive.com> (9 June 2009); "Flying Snakes Here: Demonstrate Skill," *New York Times*, 23 Aug. 1934, ProQuest Historical Newspapers.

7. "Enter U.S. Sight Unseen," *New York Times*, 24 April 1941, ProQuest Historical Newspapers; "Rare Sea Snakes Here from Tropics," *New York Times*, 9 Aug. 1935, ProQuest Historical Newspapers; "Staten Island Zoo Gets 8 New Snake Specimens," *New York Times*, 29 July 1938, ProQuest Historical Newspapers; "Sidelights of the Week," *New York Times*, 27 April 1941, ProQuest Historical Newspapers; Carl Kauffeld, *Snakes: The Keeper and the Kept*, 83–86; "Risking Death to Restore a Vicious Cobra's Vision," *Hamilton Evening Journal*, November 19, 1932, <http://www.newspaperarchive.com> (8 June 2009).

8. "Dale Carnegie's 5 Minute Biographies," *Vindicator and Republican* (Estherville Iowa) 29 March 1938, <http://www.newspaperarchive.com> (19 April 2009); Donna Risher, "Ditmars to Hunt Snakes with a Stick and a String," 3 March 1931, no source; Associated Press, "Snake Expert Invites all to Ozarks," no date or source, small news-clipping, St. Louis Zoo Records, 1910–1940, Roll 6, WHMC.

9. Lucile Mann, *From Jungle to Zoo*, 43, 42–43, 45.

10. William M. Mann, "A New Home for Reptiles," CBS Radio Address, 19 Dec.1930, Reptiles, Box 2, RU 7293, Smithsonian Institution Archive (hereafter SIA).

11. ibid. The "friend" was probably Martin Crimmins, a colonel in the U.S. Army who became a central actor in efforts to being antivenin to the United States.

12. ibid.

13. Carl Hubbs to Hon. Robert Simmons, 9 June 1930, RU 74, Box 225, F 3;

John Millen to William Mann, 29 May 1938; Harold Hermann to William Mann, 26 May 1936; Ed Bean to William Mann, 13 Dec. 1935; all in RU 74, Box 225, F 3, SIA.

14. Lucile Mann, *From Jungle to Zoo*, 109.

15. "Reptile House Repairs," Folder 7, Box 215, RU 74; William Mann to Howard F. Clark, 5 Jan. 1933, Folder 7, Box 215, RU 74; William Mann to David Ruml, 5 Jan. 1934, Folder 7, Box 215, RU 74; William Mann to David Ruml, 22 Dec. 1933, Box 215, RU 74; L.G.M. to William Mann, 2 Jan. 1934, Folder 7, Box 215, RU 74; "Memorandum to Dr. Mann," 29 Nov. 1933, Folder 7, Box 215, RU 74, SIA.

16. Rodney Dutcher, "On the Inside at Washington," *The Olean Times-Herald*, 5 Feb. 1932, <http://www.newspaperarchive.com> (9 June 2009).

17. Roger Conant and Joseph T. Collins, *Peterson Field Guide to Reptiles and Amphibians Eastern/Central North America*; Raymond Ditmars, "Our Poisonous Snakes," no source or date, St. Louis Zoo Records, 1910–1940, Roll 6, WHMCUSTA; Raymond Ditmars, "Snakes," *Field and Stream*, Jan. 1927, St. Louis Zoo Records, 1910–1940, Roll 6, WHMC.

18. Untitled press release, no date (probably 1930s), Audubon Park Commission Records, 56–7 Correspondence, Earl K. Long Library, University of New Orleans.

19. Raymond Ditmars, "Snakes," *Field and Stream*, Jan. 1927, St. Louis Zoo Records, 1910–1940, Roll 6, WHMC.

20. "Beware of Snake Bite Warns World's Greatest Serpent Hunter," Interview with Raymond Ditmars, no date or source, St. Louis Zoo Records, 1910–1940, Roll 6. WHMC; "Curious Discoveries About Poisonous Snakes in America," *San Antonio Light*, Sunday, 4 Dec.1927, <http://www.newspaperarchives.com> (5 June 2009).

21. "Deaths from Snake Bites," *New York Times*, 11 Aug. 1895, 9; M. L. Crimmins, "Poisonous Snakes and the Anti-Venin Treatment," *Southern Medical Journal*, vol. XXII, no. 7, 603–608; Afrânio do Amara, "The Snake Bite Problem in the United States and in Central America," *Bulletin of the Antivenin Institute of America*, vol. 1, no. 2 (July 1927): 31–35, 32; Henry M. Parrish, *Poisonous Snakebites in the United States*,1–10.

22. "An Epidemic of Snakes," *New York Times*, 24 Aug. 1890, p. 20; "Bitten by a Snake," *New York Times*, 8 July 1887, p. 2; "Killed by a Snake Bite," *New York Times*, 29 Sept. 1886, p. 3; "Fatally Bitten By a Snake," *New York Times*, 3 Aug. 1889, p. 2; "Bitten by a Snake," *New York Times*, 1 Aug. 1888, p. 3.

23. "The Effect of a Snake Bite," *New York Times*, 14 July 1886, p 3.

24. "Large Snakes Killed by Women. Some Club or Stone the Reptiles. Others Behead Them," *New York Times*, 9 Aug. 1894, p. 12; "Snakes for Barrooms. How a Connecticut Genius Finds a Living by Hunting Serpents," *New York Times*, 24 March 1886, p. 4; "Curious Discoveries About Poisonous Snakes in America," *San Antonio Light*, Sunday 4 Dec. 1927, <http://www.newspaperarchive.com> (5 June 2009).

25. Defying Death to Save a Crazy Cobra," *The Charleston Daily Mail*, Sunday Morning, 27 Nov. 1921, <http://www.newspaperarchive.com> (8 June 2009).

26. "Risking Death to Restore a Vicious Cobra's Vision," *Hamilton Evening Journal*, November 19, 1932, <http://www.newspaperarchive.com> (8 June 2009).

27. "Zoo Snake Breaks Away, Bites D.C. Scientist," 20 June 1927, no pages, Newspaper Clippings, no pages, no date, Box 1, Vivarium Society, SIA.

28. Raymond L. Ditmars, *Thrills of a Naturalist's Quest*, 46–56.

29. "Heroes of the Zoo," no Source or date, St. Louis Zoo Records, 1910–1940, Roll 6, WHMC; *San Antonio Light*, Thursday, 12 March 1931; *Abilene Daily*, Thursday, 12 March 1931; *Amarillo Globe*, Wednesday 4 March 1931, <http://www.newspaperarchive.com> (8 June 2009).

30. "For Rattlesnake Bite, First Chicken and then the Snake Applied to the Wound," *New York Times*, 29 July 1885, p. 2; "How They Cured a Snake Bite," *New York Times*, 4 July 1895.

31. Afrânio do Amaral, *A General Consideration of Snake Poisoning and Observations on Neotropical Pit-Vipers* (Cambridge: Harvard University Press, 1925), 19.

32. In Columbia, electric shock from a car battery was touted during the 1980s as a reliable treatment method (Ray Pawley, personal communication).

33. "Poisonous Snake Bites. Result of Dr. Allen's Experiment Made Known," *New York Times*, 24 April 1882, p. 17; "The Snake Bite Case, William Gore Very Low, But Whiskey May Save Him," *New York Times*, 16 May 1888, p 3.

34. M. L. Crimmins, "Snake-Bites and the Saving of Human Life," *Military Surgeon*, vol. 74, no. 3, March 1934, pp. 125–32; these studies are described in Crimmins (1934) and Parrish, pp. 383–85, Ray Pawley, personal communication.

35. Cited in Parrish, p. 390; C. H. Curran and Carl Kauffeld, 42.

36. Jenks Cameron, 171–185; "Snakes," Raymond Ditmars, no source or date, St. Louis Zoo Records, 1910–1940, Roll 6, WHMC.

37. Afrânio do Amaral, "The Snake Bite Problem in the United States and in Central America," *Bulletin of the Antivenin Institute of America*, vol. 1, no. 2 (July 1927): 31–35, 32; "Whiskey Saves a Snake-Bitten Horse," *New York Times*, 2 Aug. 1896, p. 2.

38. Afrânio do Amaral, "The Anti-Snake-Bite Campaign in Texas and the Sub-Tropical United States," *Bulletin of the Antivenin Institute of America*, (October 1927): 77–85, 77.

39. Genevieve Smith, "Serum for Snake Bites Repaid in Heroic Search," *New York Times*, 24 Sept. 1933, p. 109; Martin Lalor Crimmins, "Poisonous Snakes and the Antivenin Treatment," *Southern Medical Journal*, vol. XXII, no. 7 (July 1929): 603–605, 604; Martin Lalor Crimmins, "Snake Bites and the Saving of Human Life," *The Military Surgeon*, vol. 73, no. 3 (1934): 125–132, 126.

40. Doris Cochran, Meeting Minutes, 15 Jan. 1926, Box 1, Vivarium Society, RU 7163, SIA.

41. Oswaldo Vital Brazil, "History of the Primordia of Snake-Bite Accident Serotherapy," 9; Afrânio do Amaral, *Serpentes em Crise: á luz de uma legítima defensa no "caso do Butantan*," (São Paulo: 1941), 167.

42. "Prizes for 3 Scientists," *New York Times*, 28 Oct. 1927; Afrânio do Amaral, "The Brazilian Contribution Towards the Improvement of the Specific Snake Bite Treatment," *Proceedings of the New York Pathological Society*, 1923, 89–98, 90, 91.

43. Afrânio do Amaral, "The Brazilian Contribution Towards the Improvement of the Specific Snake Bite Treatment," 91.

44. "Snakes." *Time*, 28 Jan. 1929, <http://

www.time.com/time/magazine/article/0.9171.723579.00.html> (28 Feb. 2008).

45. Afrânio do Amaral, "The Snake Bite Problem in the United States and in Central America," *Bulletin of the Antivenin Institute of America*, vol. 1, no. 2 (July 1927): 31–35; Roger Conant, *A Field Guide to the Life and Times of Roger Conant*, 450–51; Pamela Henson, Lucile Quarry Mann Transcript, June–August 1977, Box 1 RU 9513, pp. 70–71, SIA.

46. Afrânio do Amaral, "The Snake Bite Problem in the United States and in Central America," *Bulletin of the Antivenin Institute of America*, vol. 1, no. 2 (July 1927): 31–35, 32; Afrânio do Amaral, "Campanhas Anti-Ophidicas," in *Memorias do Instituto Butantan*, p. 227.

47. Genevieve Smith, "Serum or Snake Bites Repaid a Heroic Search," *New York Times*, 24 Sept. 1933; "Curious Discoveries About Poisonous Snakes in America," *San Antonio Light*, Sunday, 4 Dec. 1927, <http://www.newspaperarchives.com> (5 June 2009).

48. "Beware of Snake Bite Warns World's Greatest Serpent Hunter," interview with Raymond Ditmars, no date or source, St. Louis Zoo Records, 1910–1940, Roll 6, WHMC.

49. "Serum Supply Coming to S.A.," *San Antonio Light*, Friday, 22 Oct. 1926, <http://www.newspaperarchive.com> (15 April 2009).

50. "Beware of Snake Bite Warns World's Greatest Serpent Hunter."

51. ibid; Raymond L. Ditmars, *Strange Animals I Have Known*, 194.

52. "Zoo Here Substation of Antivenin Society," 8 Dec. 1927, no source, St. Louis Zoo Records 1910—1940, Roll 6, WHMC.

53. Raymond Ditmars, *Thrills of a Naturalist's Quest*, 92.

54. Harry Goldberg, "Robbing the Snakes' Fangs of their Poison," *Fresno Bee*, 14 Aug. 1927, <http://www.newspaperarchives.com> (5 June 2009).

55. Edward Marshall, "Beware of Snake Bite Warns World's Greatest Serpent Expert," no source or date, St. Louis Zoological Park Records, 1910–1940, Roll 6, WHMC.

56. ibid.

57. ibid.

58. Martin Lalor Crimmins, "Snake Bites and the Saving of Human Life," *The Military Surgeon*, vol. 73, no. 3 (1934): 125–132, 127; Raymond L. Ditmars, *Strange Animals I Have Known*, 210; Genevieve Smith, "Serum for Snake Bites Repaid in a Heroic Search," *New York Times*, 24 Sept. 1933, p. 109.

59. Afrânio do Amaral, "Campanhas Anti-Ophidicas," in *Memorias do Instituto Butantan*, p. 230; Harry Goldberg, "Robbing the Snakes' Fangs of their Poison," *San Antonio Bee*, 14 Aug. 1927, <http://www.newspaperarchives.com> (5 June 2009); Afrânio do Amaral, p. 231; Harry Goldberg; Genevieve Smith, 1933.

60. Martin Lalor Crimmins, "Snake Bites and the Saving of Human Life," *The Military Surgeon*, vol. 73, no. 3 (1934): 125–132, 127; Dudley Jackson, 605–607, 606.

61. Afrânio do Amaral, "The Brazilian Contribution Towards the Improvement of the Specific Snake Bite Treatment," *Proceedings of the New York Pathological Society*, (1923), 89–98, 96; Dudley Jackson, 606.

62. Staten Island Zoological Society, "Snake Bite," *Animaland*, vol. XI, no. 3 (July–August, 1944): 3–4, Staten Island Zoological Society Archives.

63. Roger Conant, *A Field Guide to the Life and Times of Roger Conant*, 17–19.

64. Henry M. Parrish, *Poisonous Snakebites in the United States*, 22–23.

65. "Need of Antivenin Stations Throughout Texas," *San Antonio Express*, Friday, 16 Dec. 1927, <http://www.newspaperarchive.com> (15 April 2009; Theda Skocpol, et al., *The Politics of Social Policy in the United States*; Michael B. Katz, 133; Theda Skocpol, *Protecting Soldiers and Mothers: The Political Origins of Social Policy in the United States*.

66. Carl F. Kauffeld, "Don't Tread on Me," *News Bulletin of the Staten Island Zoological Society, Inc.*, vol. VII, no. 5 (September 1940): no pages, Staten Island Zoological Society Archives.

67. "Topics of the Times," *New York Times*, 4 Feb. 1942, p. 18.

Chapter 5

1. "Snakes at Zoo, But Living Birds are not on Menu," 1926, no source, St. Louis Zoo Records, 1910–1940, Sl 194, St. Louis Zoological Park Records; "Snakes in Zoo on Rodents and Rats," 12 Oct. 1926, no source, St. Louis Zoo Records, 1910–1940, Sl 194, St. Louis Zoological Park Records; Mrs. Charles J.

Dolan, "Snakes or Horses," *St. Louis Star*, no date, St. Louis Zoo Records, 1910–1940, SI 194, St. Louis Zoological Park Records; "Snakes in Zoo on Rodents and Rats," 12 Oct. 1926, no source, St. Louis Zoo Records, 1910–1940, SI 194, St. Louis Zoological Park Records. All articles are from the Western Historical Manuscript Collection, University of Missouri-St. Louis (hereafter WHMC).

2. Clyde Gordon, "An Appendix," *In Animaland* (May 1945): 3.

3. Earl Chapin May, "To the Zoos Has Come a New Deal," *New York Times* 10 Feb. 1935, ProQuest Historical Newspapers; "4 Lion Cubs Born in Prospect Park," *New York Times*, 5 May 1936, ProQuest Historical Newspapers.

4. For the Animal Welfare Act's regulation of zoos see *Jesse Donahue and Erik Trump, The Politics of Zoos: Exotic Animals and Their Protectors* (Dekalb: Northern Illinois University Press, 2006), 30.

5. Gus Knudson, Report to Board of Park Commissioners, 1 May 1933, p. 3, folder 51/6, Woodland Park Zoo Monthly and Annual Reports 1933–1936, Don Sherwood Parks History Collection, Seattle Municipal Archives (hereafter DSPHC). Seattle Municipal Archives will be referred to as SMA.

6. *The Report of the Department of Parks to August 1934*, pp. 27 and 148, New York City Department of Parks and Recreation Library.

7. Roger Conant, *A Field Guide to the Life and Times of Roger Conant*, 80; Marlin Perkins, *My Wild Kingdom: An Autobiography*, 80–82; quoted in Mark Rosenthal, et al., 55; Gus Knudson, "Report to Board of Park Commissioners," 1 May 1933, p. 3, folder 51/6, Woodland Park Zoo Monthly and Annual Reports 1933–1936, DSPHC; Robert Moses, "You Can Trust the Public," *The American Magazine* (July 1939): 144–45, 58–59, "disgrace" quotation from p. 59; Robert Moses, "The Public Parks and Parkways of New York City and Long Island" (unpublished transcription of a speech), 8 April 1936, p. 28, box 11, folder M, Robert Moses Papers, Manuscripts and Archives Section, New York Public Library; *The Report of the Department of Parks to August 1934*, pp. 27 and 148, New York City Department of Parks and Recreation Library; quoted in Caroline A. A. Baker, 17, Central Park Zoo file, New York City Department of Parks and Recreation Library; Eugene Kinkead and Russell Maloney, "Central Park II —'A Nasty Place,'" *The New Yorker* (20 Sept. 1940): 34–45.

8. Editorials (1922 and 1926) quoted in Caroline A. A. Baker, 17, Central Park Zoo file, New York City Department of Parks and Recreation Library; *Annual Report, Department of Parks, Manhattan* (1929), pp. 61–62, New York City Department of Parks and Recreation Library; *Annual Report of the Department of Parks, Borough of Manhattan* (1931), p. 82, New York City Department of Parks and Recreation Library; "Low Speed in Park Favored by Herrick," *New York Times*, 27 Sept. 1931, ProQuest Historical Newspapers.

9. David Hancocks, *Animals and Architecture*, 125.

10. David Ehrlinger, 42–43, 59; David Hancocks, *Animals and Architecture*, 125–31; Mark Rosenthal, et al., 53–54.

11. Ted J. Ligibel, 47–67.

12. David Ehrlinger, 72.

13. "Tiny Mischka Bites Policeman in Wild Chase After Stroll from 'Escape-Proof' Cage," *New York Times*, 21 May 1935; "Sun Bear at the Zoo Nearly Gets Mauling," *New York Times*, 6 Nov. 1935. Both accessed through ProQuest Historical Newspapers.

14. Belle Benchley, *My Friends the Apes* (Boston: Little, Brown and Company, 1942), 254; Raymond L. Ditmars, *Thrills of a Naturalist's Quest*, 188.

15. Mark Rosenthal, et al., 52–53; Pamela Henson, Lucile Quarry Mann Transcript, June–August 1977, Box 1 RU 9513, pp. 167–70, Smithsonian Institution Archives (hereafter SIA).

16. Raymond L. Ditmars, *Thrills of a Naturalist's Quest*, 188, 189; Marlin Perkins, *My Wild Kingdom: An Autobiography*, 33; David Ehrlinger, 44.

17. Raymond L. Ditmars, *Thrills of a Naturalist's Quest*, 194; Marlin Perkins, *My Wild Kingdom: An Autobiography*, 71; "Elephant Proves Memory and Greets Zoo Keeper Who Aided Her When Ill in Fall," *New York Times*, 22 June 1935, ProQuest Historical Newspapers.

18. "Lioness Claws Central Park Zoo Director As He Shows Pet to Visitor from Berlin," *New York Times*, 29 July 1936, ProQuest Historical Newspapers; "600-Pound

Pet Bear Kills a Cub, Claws 2 Others in Central Park," *New York Times*, 21 Sept. 1935, ProQuest Historical Newspaper — stories of Stout's escapades appeared frequently in the *New York Times*; *The Report of the Department of Parks to August 1934*, p. 8, New York City Department of Parks and Recreation Library; "Zoo Head Ousted at Central Park," *New York Times*, 4 April 1939, ProQuest Historical Newspapers; "irresponsible" quotation in Caroline A. A. Baker, 21, Central Park Zoo file, New York City Department of Parks and Recreation Library; Eugene Kinkead and Russell Maloney, "Central Park III — What a *Nice* Municipal Park," *The New Yorker* (27 Sept. 1940). Stout was reinstated after a prolonged court battle; see "Zoo Head Seeks Pension," *New York Times*, 14 June 1941, ProQuest Historical Newspapers; Mark Rosenthal, et al., 68.

19. Raymond L. Ditmars, *Thrills of a Naturalist's Quest*, 190.

20. Marlin Perkins, *My Wild Kingdom: An Autobiography*, 73.

21. Belle J. Benchley, *My Friends the Apes*, 109.

22. ibid, 211–212; Pascal James Imperato and Eleanor M. Imperato, 167; Marlin Perkins, *My Wild Kingdom: An Autobiography*, 77; Belle J. Benchley, *My Friends the Apes*, 212.

23. Belle J. Benchley, *My Friends the Apes* , 161.

24. ibid, 161, 125.

25. "Animal Picture Book on Sale at Zoo Today," *New York Times*, 28 June 1935, ProQuest Historical Newspapers; Myrtle Glenn Terry, *The Picture Book Zoo* (New York City Department of Parks, 1935). Swann's married name was Saarinen.

26. Franklin D. Roosevelt, *The Happy Warrior, Alfred E. Smith: A Study of a Public Servant*, 36–37.

27. W. J. Davies, "The Zoo," *Post-Intelligencer*, 10 May 1939, folder 16/5, Woodland Park and Zoo 1922–1939, Ben Evans Recreation Program History Collection, SMA.

28. Gus Knudson, "Monthly Report," 31 May 1935, folder 51/6, Woodland Park Zoo Monthly and Annual Reports 1933–1936, DSPHC; Annual Report for 1943, Board of Commissioners, Seattle Park Department Annual Reports 1936–1950 (excerpts relating to Woodland Park Zoo), Woodland Park Zoo Archives (hereafter WPZA).

29. "Ran Away From Home 32 Years Ago; Now Gus Is Going to Go Back," newspaper clipping, no source, May 1916, Knudson's Scrapbook, WPZA; "It Sounds Funny, But Snakes Are a Great Aid to Romance," *Post Intelligencer*, 6 June 1947, folder 16/6, Woodland Park and Zoo 1940–1951, Ben Evans Recreation Program History Collection, SMA; Don Sherwood, handwritten memo on Woodland Park's history, 4 Dec. 1975, WPZA.

30. Business cards in Knudson's personal scrapbook, Woodland Park Zoo Archives; Gudrun Anderson, untitled newspaper story, no source, no date, Knudson's Scrapbook, WPAZ; newspaper article by Gudrun Andersen, no source or date, Knudson's Scrapbook, WPZA; "Tiger, Under Knife, Retains Stoic Calm," *New York Times*, 22 March 1932; "Liquor Robs Camel of Symbolic Right," *New York Times*, 27 Oct. 1932, ProQuest Historical Newspapers; Gus Knudson, "Monthly Report," 30 April 1933, p. 3, folder 51/6, Woodland Park Zoo Monthly and Annual Reports 1933–1936, DSPHC; "Doctor at Park Zoo," newspaper clipping, no source, no date, Knudson's Scrapbook, WPZA.

31. *19th Annual Report of the Board of Park Commissioners of the City of Seattle for 1922*, Seattle Park Department Annual Reports 1904–1935 (excerpts relating to Woodland Park Zoo), WPZA; Gus Knudson, "Monthly Report," 31 May 1933, p. 1, folder 51/6, Woodland Park Zoo Monthly and Annual Reports 1933–1936, DSPHC.

32. Gus Knudson, "Monthly Report," 30 Sept. 1933, p. 1, folder 51/6, Woodland Park Zoo Monthly and Annual Reports 1933–1936, DSPHC; Gus Knudson to Board of Park Commissioners, 19 Nov. 1934, folder 5/4, Correspondence and Financial Records (1934), DSPHC.

33. S. B. Groff, "Lioness Slays Cub to Save it From Life of Captivity," no source, 9 June 1928, Knudson's Scrapbook, 1910s–1930s, WPZA.

34. "Park Board Orders Investigation of Zoo at Woodland," newspaper clipping, 27 Dec. 1926, Miscellaneous Files; "New Zoo Advised by Humane Body," *Seattle Times*, 24 March 1927, Miscellaneous Files; "Humane Society Seeks Improvement for Zoo," *Seattle Times*, 1 April 1927, Miscellaneous Files; "Council to Vote on Woodland Park Zoo," *Seattle Times*, 10 May 1927,

Miscellaneous Files. All of the articles are from the WPZA.

35. "Humane Society Finds All Animals Living in Deplorable State — Zoological Organization Proposed — Park Chief Opposes Removal," *Seattle Times*, 7 July 1927, Miscellaneous Files, WPZA.

36. ibid.

37. Board of Park Commissioners to Mayor and City Council, 31 March 1930, p. 13, folder 4/1, Correspondence and Financial Records, DSPHC, SMA; *Report of the Board of Park Commissioners of the City of Seattle 1923–1930*, p. 79, folder 4/4, Annual Report 1923–30, DSPH, SMA. Emphasis added.

38. Regular Meeting Minutes, Seattle Board of Park Commissioners, 8 May 1930, Woodland Park Zoo Archives; Gus Knudson, typescript report of zoo tour, June 1930, pp. 1 and 24–25, folder 51/8, Woodland Park Zoo Reports on Tours of Other 1940–1948, DSPHC, SMA; *Report of the Board of Park Commissioners of the City of Seattle 1923–1930*, p. 79, folder 4/4, Annual Report 1923–30, DSPHC, SMA; H. E. Hoffman, "A Ten Year Program for the Seattle Park Department," 25 May 1931, pp. 88–89, folder 4/11, Ten Year Program for the Seattle Park Department 1931, DSPHC, SMA.

39. Quoted in Park Engineer to Ira C. Brown, 31 Dec. 1930, folder 50/5, Woodland Park Zoo Animals — Popular/Totem 1916–1949, DSPHC, SMA.

40. See San Diego Zoo list of animals for sale, Audubon Park Commission Records, Box 56–8 Audubon Park Commission 1923–1927, Earl K. Long Library, University of New Orleans.

41. Elizabeth Hanson, 44.

42. Park Board Commission Records, 8 Sept. 1932, WPZA.

43. Park Board Commission Records, 29 Dec. 1932 and 2 Feb. 1933, WPZA.

44. "Camel Brings $600 at 'Clearance Sale,'" *New York Times*, 1 July 1930, ProQuest Historical Newspapers; "Animals Go Begging at Park Auction," *New York Times*, 23 June 1932, ProQuest Historical Newspapers; Park Board Commission Records, 22 Dec. 1932, 26 Jan. 1933, and 9 Feb. 1933, WPZA.

45. Park Board Commission Records, 8 March 1933, WPZA.

46. "Recapitulation of Animals in the Zoological Collection, 12–31–39," folder 5/21, Zoo Inventory (1939); Knudson mentioned other zoos' slaughterhouses in his typescript report of his zoo tour, June 1930, p. 26, folder 51/8, Woodland Park Zoo Reports on Tours of Other 1940–1948; Gus Knudson, "Zoo Report for 1933," folder 51/6, Woodland Park Zoo Monthly and Annual Reports 1933–1936; Gus Knudson, "Monthly Report," 30 April 1933, p. 3, folder 51/6, Woodland Park Zoo Monthly and Annual Reports 1933–1936; Gus Knudson, "Monthly Report," 31 Aug. 1933, p. 2, folder 51/6, Woodland Park Zoo Monthly and Annual Reports 1933–1936. All materials are from the DSPHC, SMA.

47. Gus Knudson, "Monthly Report," 31 Oct. 1933, p. 3, folder 51/6, Woodland Park Zoo Monthly and Annual Reports 1933–1936, DSPHC, SMA.

48. Gus Knudson, "Zoo Report for 1933," folder 51/6, Woodland Park Zoo Monthly and Annual Reports 1933–1936, DSPHC, SMA; Minutes of Audubon Park Commission, 15 Oct. 1941, Audubon Park Commission Records, 56–142, Minutes of APC 1940–41, Earl K. Long Library, University of New Orleans; Park Board Commission Records, 26 Sept. 1935, WPZA.

49. R. D. Lelian to Seattle City Council, City Commissioner of Parks, and Park Board, 3 Sept. 1935, CFN # 148070, Documents from City Council Petitions, Seattle Municipal Archives. Lelian is elsewhere referred to as Lehman, although the former appears to be the correct spelling.

50. "Woodland Park Zoo Animals Well Treated, Says Knudson," no source or date, Knudson's Scrapbook, WPZA.

51. Chris Christenson to City Council, 16 Sept. 1935, CFN # 148070, Documents from City Council Petitions, SMA.

52. Harry Westfall (President, Board of Park Commissioners) to City Council, 26 Sept. 1935, CFN # 148070, Documents from City Council Petitions, SMA.

53. Redick H. McKee and Mabel McGill to Park Commissioners, 30 Dec. 1935, CFN # 149229, Documents from City Council Petitions, SMA. A copy with accompanying note was sent to the City Council at the same time, Park Board Commission Records, 2 Jan. 1936 and 16 Jan. 1936, WPZA; Gus Knudson, "Monthly Report," 31 Jan. 1936, p. 3, folder 51/6, Woodland Park Zoo Monthly

and Annual Reports 1933–1936, DSPHC, SMA.

54. Gus Knudson, "Air Raid Defense at the Woodland Park Zoological Gardens," 1942, folder 50/1, Woodland Park and Zoo 1906–1949, DSPHC, SMA.

55. Elizabeth Hanson, 81, 83.

56. George Lewis and Byron Fish, 50. Sammy, a rogue bull elephant donated to the Detroit Zoo, was soon put down.

57. Tusko's career is chronicled in George Lewis and Byron Fish, *I Loved Rogues*.

58. David Hancocks, *Woodland Park Zoological Gardens Seventy-Fifth Anniversary 1904–1979* (Zoo Foundation of Woodland Park Zoological Gardens, 1979), 12; "World's largest" and "poor beast" quotations from Gus Knudson, Report to Board of Park Commissioners, 1 May 1933, p. 1, folder 51/6, Woodland Park Zoo Monthly and Annual Reports 1933–1936, DSPHC, SMA; 8 Oct. 1932 and 28 Nov. 1932, Woodland Park Zoo Archives; crowd estimate from Gus Knudson to Board of Park Commissioners, 11 Oct. 1932, Parks Central Reference File, "Tusko 1932–1940," SMA.

59. Minutes of Board of Park Commissioners, 15 Sept. 1932 and 22 Sept. 1932, Woodland Park Zoo Archives.

60. Knudson noted that Tusko was an "old friend" of S. Lasher, a reporter for the American Association of Zoological Parks and Aquariums, who had "been keeping track of the big elephant for years," Gus Knudson, "Monthly Report," 31 May 1933, p. 1, folder 51/6, Woodland Park Zoo Monthly and Annual Reports 1933–1936, DSPHC, Seattle Municipal Archives; "It Sounds Funny, But Snakes Are a Great Aid to Romance," *Post Intelligencer*, 6 June 1947, folder 16/6, Woodland Park and Zoo 1940–1951, Ben Evans Recreation Program History Collection, SMA. By "bullets," Knudson probably meant "buckshot." Tusko never killed anyone, according to Homer C. Walton, "The M.L. Clark Wagon Show," *Bandwagon*, vol. 9, no. 2 (March–April, 1965): 4–11, <http://www.circushistory.org/Bandwagon/bw-1965Mar.htm>; George Lewis and Byron Fish, 42.

61. Gus Knudson, Report to Board of Park Commissioners, 1 May 1933, p. 1, folder 51/6, Woodland Park Zoo Monthly and Annual Reports 1933–1936; Gus Knudson, Monthly Report, 30 April 1933, p. 1, folder 51/6, Woodland Park Zoo Monthly and Annual Reports 1933–1936. Both reports are from the DSPHC, SMA.

62. Gus Knudson, Report to Board of Park Commissioners, 1 May 1933, p. 1, folder 51/6, Woodland Park Zoo Monthly and Annual Reports 1933–1936, DSPHC, SMA; B.B. Lustig to Mayor John F. Dore, 12 Jan. 1933, Parks Central Reference File, "Tusko 1932–1940," SMA; Minutes of Board of Park Commissioners, 16 Feb. 1933, WPZA; Gus Knudson, Monthly Report, 30 April 1933, p. 1, folder 51/6, Woodland Park Zoo Monthly and Annual Reports 1933–1936, DSPHC, SMA.

63. B.B. Lustig to Mayor John F. Dore, 12 Jan. 1933, Parks Central Reference File, "Tusko 1932–1940," SMA.

64. Park Board Commission Records, 12 Sept. 1933, WPZA; Park Board Commission Records, 22 March 1933, WPZA; Gus Knudson, Report to Board of Park Commissioners, 1 May 1933, p. 1, folder 51/6, Woodland Park Zoo Monthly and Annual Reports 1933–1936, DSPHC, SMA; Gus Knudson, Monthly Report, 30 April 1933, p. 1, folder 51/6, Woodland Park Zoo Monthly and Annual Reports 1933–1936, DSPHC, SMA; Gus Knudson, Report to Board of Park Commissioners, 1 May 1933, p. 2, folder 51/6, Woodland Park Zoo Monthly and Annual Reports 1933–1936, DSPHC, SMA.

65. Park Board Commission Records, 17 May 1933 and 24 May 1933, WPZA; Gus Knudson to Board of Park Commissioners, 12 June 1933, Parks Central Reference File, "Tusko 1932–1940," SMA; Park Board Commission Records, 14 June 1933, WPZA.

66. Gus Knudson, "Monthly Report," 30 June 1933, p. 4, folder 51/6, Woodland Park Zoo Monthly and Annual Reports 1933–1936, DSPHC, SMA.

67. Gus Knudson, "Monthly Report," 31 Aug. 1933, p. 3, folder 51/6, Woodland Park Zoo Monthly and Annual Reports 1933–1936, DSPHC, SMA.

68. Gus Knudson, "Monthly Report," 31 Oct. 1933, p. 1, folder 51/6, Woodland Park Zoo Monthly and Annual Reports 1933–1936; Gus Knudson, "Monthly Report," 30 Nov. 1933, p. 3, folder 51/6, Woodland Park Zoo Monthly and Annual Reports 1933–1936; Gus Knudson, "Monthly Report," 30 April 1934, p. 3, folder 51/6, Woodland Park Zoo Monthly and Annual Reports 1933–

1936; Gus Knudson, "Monthly Report," 28 Feb. 1935, p. 4, folder 51/6, Woodland Park Zoo Monthly and Annual Reports 1933–1936. All documents are from the DSPHC, SMA.

69. "Mayor Backs Move to Drop Woodland Park Zoo," newspaper clipping, no source or date, Woodland Park Zoological Gardens Chronological List of Projects, WPZA.

70. ibid.

71. Gus Knudson, "Report of Woodland Park Zoo for 1934," pp. 1–2, folder 51/6, Woodland Park Zoo Monthly and Annual Reports 1933–1936; Minutes of Board of Park Commissioners, 10 Jan. 1934 and 24 Jan. 1934, WPZA; Gus Knudson, "Improvements at the Woodland Park Zoological Gardens 1940," folder 5/23, Construction and Maintenance (1940). Both documents are from the DSPHC, SMA.

72. Gus Knudson, "1934 Zoo Report," pp. 1–3, 4–5, folder 51/6, Woodland Park Zoo Monthly and Annual Reports 1933–1936, DSPHC, SMA.

73. ibid, p. 6.

74. ibid; Gus Knudson, "Monthly Report," 30 Sept. 1934, p. 4, folder 51/6, Woodland Park Zoo Monthly and Annual Reports 1933–1936, DSPHC, SMA.

75. Gus Knudson, "Monthly Report," 31 Aug. 1935, p. 4, folder 51/6, Woodland Park Zoo Monthly and Annual Reports 1933–1936, DSPHC, SMA.

76. A. E. Schutt (Park Commissioner) to the Editor, 4 March 1935, Folder 5/7, Correspondence and Financial Records (1935); James A. Wood, "Speaking for the Times" (editorial), *Seattle Times*, 26 April 1936, folder 5/10, Correspondence and Financial Records; "Changes Asked for Park Board," *Seattle Times*, 26 April 1936, folder 5/10, Correspondence and Financial Records. All of the documents are from the DSPHC, SMA.

77. For details on WPA work, see Seattle Park Department Annual Reports 1936–1950 (Excerpts Relating to Woodland Park Zoo), WPZA; Redick H. McKee and Mabel V. McGill to Seattle City Council, 1 Sept. 1936, CFN# 152279, Documents from City Council Petitions, SMA; Harbor and Public Grounds Committee to Mabel V. McGill, no date, CFN# 152279, Documents from City Council Petitions, SMA. The city suggested that the project would be a "splendid memorial to the work of the association"; Gus Knudson, "Betterments Made at the Zoological Gardens 1939: W. P. A. Work," p. 2, folder 5/17, Construction and Maintenance (1939), DSPHC, SMA; Redick H. McKee and Mabel V. McGill to Seattle City Council, 1 Sept. 1936, CFN# 152279, Documents from City Council Petitions, SMA.

78. Prices are from Henry Bartels (animal trader) to New Orleans Zoo Society (27 April 1937), and from a Henry Trefflich price list (no date — probably 1941). Both in Audubon Park Commission Records, Box 56–9 Correspondence, Earl K. Long Library, University of New Orleans.

79. Petition, 8 July 1938, CFN # 159829, Documents from City Council Petitions, SMA; Knudson formally proposed the island, with plans and cost estimates, on Nov. 26, 1937, Regular Meeting Minutes, Seattle Board of Park Commissioners, 27 Oct. 1938, WPZA; 1940 Annual Report of Board of Park Commissioners, Seattle Park Department Annual Reports 1936–1950 (excerpts relating to Woodland Park Zoo), WPZA; "Praise for Park Board," *Post-Intelligencer*, Aug. 1940, folder 16/6, Woodland Park and Zoo 1940–1951, Ben Evans Recreation Program History Collection, SMA; "Monkeyshines," newspaper clipping, no source, 21 Aug. 1940, folder 16/6, Woodland Park and Zoo 1940–1951, Ben Evans Recreation Program History Collection, SMA.

80. "Monkey School Board Meets in Peace, Departs in Pieces," no date, author or source, SMA 16/6, newspaper clippings, WPZA.

81. Doug Welch, "Monkey Isle Revolt Still Simmers," *Seattle Post Intelligence*, 20 Aug. 1940, folder 16/6, SMA; "Monkeyshines," newspaper clipping, no source, 21 Aug. 1940, folder 16/6, SMA.

82. "Principals in Monkey Island War," no author, date, or source, folder 16/6, SMA.

83. Forrest Williams, "Insurgents Heave Reactionary King Into Moat, Crown Suffragist," newspaper clipping, 2 June 1940; "Principals in Monkey Island War," no author, date, or source; Dough Welch, "Sing Cows Simian Colony Into Submission — For a Day," newspaper clipping, 16 Aug. 1940; Doug Welch, "Monkey Isle Revolt Still Simmers," *Seattle Post Intelligence*, 20 Aug. 1940; "'Bun's-Rush' Likely for Yuko in Monkey Isle 'Voting'

Today," *Seattle Times*, 25 Aug. 1940; Matthew O'Connor, "Sing Dynasty on Monkey Isle Ends; Dictator Gets Dunking," no date or source. All clippings in folder 16/6, SMA.

84. Park Board Commission Records, 7 Aug. 1938, Woodland Park Zoo Archives; "Status of Projects: Seattle Parks and Playfields and Laurelhurst Playfield," 29 July 1940, folder 63/3, WPA Projects 1935–1941, DSPHC, Seattle Municipal Archives; Park Board Commission Records, 16 Oct. 1941 and 16 April 1942, Woodland Park Zoo Archives; Laura B. Rawsthorn to Councilman James Scavotto, 4 Aug. 1942, CFN # 174775, Documents from City Council Petitions, Seattle Municipal Archives; Board of Park Commissioners to City Council, 21 Aug. 1942, CFN # 174775, Documents from City Council Petitions, SMA.

85. Gus Knudson to Board of Park Commissioners, 29 Jan. 1947, folder 50/1, Woodland Park and Zoo 1906–1949, DSPHC, SMA; Don Magnuson, "Zoo Director, Retiring, Raps Park Operation," *Seattle Times*, 21 March 1947, folder 16/6, Woodland Park and Zoo 1940–1951, Ben Evans Recreation Program History Collection, SMA; "Knudson Raps Woodland Zoo as Disgrace," no source or date, clipping from Knudson's Scrapbook, Woodland Park Zoo Archives; Paul V. Brown, "Seattle's Shabby, Inadequate Zoo," *Seattle Times*, 27 May 1951, folder 16/7, Woodland Park and Zoo 1950–1952, Ben Evans Recreation Program History Collection, Seattle Municipal Archives; "Knudson Raps Woodland Zoo as Disgrace," no source or date, clipping from Knudson's Scrapbook, WPZA.

Chapter 6

1. Pamela Henson, Lucile Quarry Mann Transcript, June–August 1977, Box 1 RU 9513, p. 42, Smithsonian Institution Archives (hereafter SIA).

2. For a similar finding of how wars hurt animals see Deanne Stillman, *Mustang: The Saga of the Wild Horse in the American West.*

3. Gus Knudson, "Air Raid Defense at the Woodland Park Zoological Gardens," 1942, folder 50/1, Woodland Park and Zoo 1906–1949, Don Sherwood Parks History Collection, Seattle Municipal Archives (hereafter DSPHC, SMA); Gus Knudson to All Keepers at the Zoo, 15 Dec. 1941, folder 6/20, DSPHC, SMA.

4. "War Streamlines Tigers at Zoo, But Monkeys Thrive on Ersatz," *New York Times*, 10 Jan. 1943, ProQuest Historical Newspapers; Minutes of Audubon Park Commission, 16 Dec. 1942 and 2 April 1943, Audubon Park Commission Records, 56–142, Minutes of APC 1942 and 1943, Earl K. Long Library, University of New Orleans; Stuart Whitehouse, "Zoo Curator 'Rages' At Hay Ration Rumor," *Seattle Star*, 8 Oct. 1943, folder 16/6, Woodland Park and Zoo 1940–1951, Ben Evans Recreation Program History Collection, SMA; Minutes of New Orleans Zoological Society, 15 Sept. 1941, Audubon Park Commission Records, 56–155, Minutes of the New Orleans Zoological Society July 1941–September 1942, Earl K. Long Library, University of New Orleans.

5. Al Vida to Mr. Mitchell, 29 Aug. 1940, Office of General Director, Allyn Jennings, Wildlife Conservation Society Archives, New York; "3 Gorillas Added to Zoo at Central Park; Given to the City by Anonymous Donor," *New York Times*, 2 July 1943, ProQuest Historical Newspapers; "Zoos, Facing Animal Shortage, May Set Up A Beast 'Ellis Island' in East Africa," *New York Times*, 13 July 1941, ProQuest Historical Newspapers.

6. Marlin Perkins, *My Wild Kingdom: An Autobiography*, 87; Thos. W. Kracke to New Orleans Zoological Society, 31 March 1941, Audubon Park Commission Records, Box 56–15 Correspondence, Earl K. Long Library, University of New Orleans.

7. Gus Knudson, "Report on Betterments, Repairs, and Miscellaneous Work Accomplished by the WPA at the Woodland Park Zoological Gardens during the year 1941," 1942, pp. 1–7, folder 6/6, Correspondence and Financial Records 1941, DSPHC, SMA; George Lewis and Byron Fish, 109; Gus Knudson, "Report on Betterments, Repairs, and Miscellaneous Work"; Minutes of Audubon Park Commission, 17 Feb. and 18 Aug. 1943, Audubon Park Commission Records, 56–142, Minutes of APC 1943, Earl K. Long Library, University of New Orleans.

8. For an example of how social welfare benefits have gone to soldiers first see Theda Skocpol, *Protecting Soldiers and Mothers: The Political Origins of Social Policy in the United States*; Rich Buickerood, 11, 13–14, quotation from p. 11.

9. Robert Moses, *Public Works: A Dangerous Trade*, 697.

10. Jonathan Kuhn, "Central Park Wildlife Conservation Center," March 2001, Central Park Zoo file, New York City Department of Parks and Recreation Library; Robert Moses, *Public Works: A Dangerous Trade*, 28; New York Zoological Society and Jerry M. Johnson, Inc., *A Proposal for the Renovation of Prospect Park Zoo* (no date—1980s), 6–8, 10, Prospect Park Archives.

11. "A Future for the City's Zoos," (no date—1980?), Central Park Zoo file, New York City Department of Parks and Recreation Library; New York Zoological Society and Jerry M. Johnson, Inc., *A Proposal for the Renovation of Prospect Park Zoo* (no date—1980s), 8, 10–12, 15Prospect Park Archives.

12. David Ehrlinger, 69; Ohio Writer's Project, *Guide Book: The Cincinnati Zoo*, 89; David Kleck, "The Angry Animals," *New Orleans Magazine* (1968), quoted in Bureau of Governmental Research, *Audubon Park Zoo Study: Part I—Zoo Improvement Plan* New Orleans, 1971), 1; Bureau of Governmental Research, *Audubon Park Zoo Study: Part I—Zoo Improvement Plan* New Orleans, 1971), 1–3.

13. Jack Boulware, "Goodbye to Monkey Island," Slap Shots, *SF Weekly*, 19 July 1995, <http://http://www.sfweekly.com/1995-07-19/calendar/slap-shots/print> (28 June 2008); G. R. Bartholick (architect-planner), "Long Range Master Plan Study—Woodland Park and Seattle Zoo" (1970), p. C-12, folder 51/4, Woodland Park Zoo—Long Range Master Plan Study, DSPHC, SMA; L. Ronald Forman, 134.

14. Jesse Donahue and Erik Trump, *The Politics of Zoos: Exotic Animals and Their Protectors* (DeKalb, IL: Northern Illinois University Press, 2006).

15. The relationship between increased government rules and the cost of doing business has been noted by many, including James Q. Wilson, *Bureaucracy: What Government Agencies Do and Why They Do It*, 342–44.

16. Senate Subcommittee on the Smithsonian Institution of the Committee on Rules and Administration, *Federal Assistance for Zoos and Aquariums: Hearing before the Subcommittee on the Smithsonian Institution of the Committee on Rules and Administration*, 93rd Cong., 2nd sess., 1974, 3–23; quotations from pp. 22–23. Two versions of the same bill (S. 2042 and S. 2774) were under consideration, and Representative G. William Whitehurst had introduced a similar bill in the House.

17. Pamela Henson, Lucile Quarry Mann Transcript, June–August 1977, Box 1 RU 9513, pp. 195–200, SIA.

18. Terry L. Maple and Erika F. Archibald, 172–73; Figure comes from 2002 AZA member surveys. For the sake of simplicity, we will use the term "privatization" throughout this chapter. Privatization advocates sometimes prefer the term "public-private partnership" because it suggests that privatized services are still fundamentally "public" in nature. This perception is particularly important for zoos, which depend on their public status to enjoy special exemptions related to traffic in wild animals. On the history of the term, see E. S. Savas, *Privatization and Public-Private Partnerships*, 3–5; Robert Bailey, "Uses and Misuses of Privatization," in *Prospects for Privatization*, ed. Steve Hanke (New York: Academy of Political Science, 1987), 138–52; Elliott Sclar, vii.

19. Most political scientists and economists who have written about privatization mention zoos in passing, if at all. Thus, we were unable to turn to them for specific arguments about this topic, but we have relied on their general arguments about the purposes of and obstacles to privatization. See the previous footnote for some privatization sources. Keith Wade, "It's a Jungle Out There! What We Can Learn from the Privatization of Zoos," *The Freeman*, vol. 48, no. 8, 2003, 464–466; David Haarmeyer and Elizabeth Larson, 121–25. The authors are associated with the Libertarian website www.libertyhaven.com.

20. The AZA has a collection of these privatization documents in an *AZA Private/Public Task Force* binder available to members (no date). This collection documents the privatization of eight zoos.

21. Jon Charles Coe, 111; Privatization case study reports on San Francisco, Pittsburgh, Lincoln Park, and Woodland Park zoos; collected in *AZA Private/Public Task Force* binder (no date, no page numbers); Privatization case study of San Francisco Zoo, collected in *AZA Private/Public Task Force* binder (no date), p. 2. E. S. Savas argues that governments decide to contract out public services

when four conditions are met: "serious fiscal stress," possibility of cost savings to government, political feasibility of privatization, and a "precipitating event" (e.g., loss of accreditation for a hospital). Clearly all of these conditions have been present in most zoo privatizations. See E. S. Savas, *Privatization: The Key to Better Government* (Chatham, NJ: Chatham, 1987), 255–57.

22. "The Pittsburgh Zoo: A Public-Private Partnership, Business Plan," (n.p., January 1993), collected in *AZA Private/Public Task Force* binder (no date); San Francisco Zoological Society "5-Year Plan: 1993–1998 San Francisco Zoological Gardens," (15 July 1993), 5, 12, collected in *AZA Private/Public Task Force* binder (no date).

23. See the Audubon Zoo for data on this; Bureau of Governmental Research, *Audubon Park Zoo Study: Part I — Zoo Improvement Plan* (New Orleans, 1971), 89.

24. Belle Benchley to Allyn Jennings, 17 Oct. 1940, Office of General Director, Allyn Jennings, Wildlife Conservation Society Archives, New York; Bureau of Governmental Research, *Audubon Park Zoo Study: Part I — Zoo Improvement Plan* (New Orleans, 1971), 89.

25. Molly Selvin, "Manuel Mollinedo," *Los Angeles Times*, 15 Oct. 2000 (retrieved from Newsbank online database 7 May 2001); "Corporate Sponsorship Bill," *Communiqué* (September 1993): 2.

26. Robert Ramin, "The New Corporate Philanthropy," *Communiqué* (August 1994): 5–6; Robert Ramin, "Development," *Communiqué* (April 1994): 2.

27. *Amusement Business*, 10 Dec. 2001, retrieved from ProQuest online database 18 Feb. 2002; Tim O'Brien, "Fort Worth Zoo Getting Wild!," *Amusement Business*, 21 May 2001, retrieved from BusIndustry online database 18 Feb. 2002; Sally Nancrede, "Indianapolis Zoo Unveils Roller Coaster in Elephant Area," *Indianapolis Star*, 19 April 2001), retrieved from BusIndustry online database 18 Feb. 2002; Bonnie Cavanaugh, "Columbus Zoo Revamps Its Dining Facilities," *Nation's Restaurant News*, 9 Aug. 1999, retrieved from ProQuest online database 18 Feb. 2002; Amber Bicknase, "Kidding with Zoo Food," *Restaurant Hospitality* 84:1 (January 2000), retrieved from ProQuest online database 18 Feb. 2002; Louis Cain and Dennis Meritt Jr., "The Growing Commercialism of Zoos and Aquariums," *Journal of Policy Analysis and Management* 17, no. 2 (1998): 298–312.

28. Nicole Mazur, 218.

29. "Zoo Commission II: A Report and Recommendations to the Mayor on the Strategic Direction of the Woodland Park Zoo," January 1996, p. 16, collected in *AZA Private/Public Task Force* binder (no date); as the zoo puts it, "In hindsight, we wished early negotiations had included water service, control of parking lots and a minimum base City subsidy rather than previous subsidy dropping to zero," Privatization case study of Zoo Atlanta, collected in *AZA Private/Public Task Force* binder (no date, no page).

30. Dorian Block, "The Bronx Zoo Funding Endangered," 7 Jan. 2009, <http://www.nydailynews.com>; Henry Fountain, "In Zoo Cuts, It's Man vs. Beast," *The New York Times*, 19 March 2009, 28; William Kates, "Zoos, aquariums face the ax in NY, elsewhere," 13 Jan. 2009, <http://www.news.yahoo.com>.

31. Dorian Block, "The Bronx Zoo Funding Endangered," 7 Jan. 2009, <http://www.nydailynews.com>; William Kates, "Zoos, Aquariums Face the Ax in NY, Elsewhere," 13 Jan. 2009, <http://www.news.yahoo.com>; Jenn Bain, "New York's living museums facing cuts," 3 Dec. 2008, <http://www.msnbc.com>.

32. "Mayors' infrastructure request full of pork, critic says," 18 Dec. 2008, <http://www.cnn.com>; "Congress looks to keep stimulus cash from zoos, golf courses," 2 Feb. 2009, <http://www.cnn.site.printthis.clickability.com>; "Zoos, aquariums face the ax in NY, elsewhere," 13 Jan. 2009,<http://www.news.yahoo.com>; "Mayors' infrastructure request full of pork, critic says," 18 Dec. 2008, <http://www.cnn.com>.

33. Dorian Block, "The Bronx Zoo funding endangered," 7 Jan. 2009, <http://www.nydailynews.com>.

34. Michael to Paul Stanton (Department of Parks and Recreation), 16 July 1980, Prospect Park Zoo file, New York City Department of Parks and Recreation Library; Joseph Bresnan to Katherine Wickham, 5 Aug. 1981, Prospect Park Zoo file, New York City Department of Parks and Recreation Library.

35. "An Artist for the Park," *Dialogue*

(Dec. 1988/Jan. 1989): 6; Jonathan Kuhn, "Central Park Wildlife Conservation Center," March 2001, Central Park Zoo file, New York City Department of Parks and Recreation Library.

36. Caroline Durieux, "Report of Federal Art Project of Louisiana as of May 1, 1939," Works Progress Administration Federal Art Project, 1934–1970, Reel 5289, Frames 1164–1169, SIA, available at <http://www.aaa.si.edu/collections/digitalcollections/collectionsonline/cahiholg>.

37. Pueblo Zoo, "Islands of Life at the Pueblo Zoo: History of the Animal House and Adjoining Monkey Island," 9 April 2007, <http://www.pueblozoo.org/news/IOL4-9-07.htm> (28 June 2008).

38. "Daily Parks Report," 10 Feb. 1984, Central Park Zoo file, New York City Department of Parks and Recreation Library; Peggy Earle, "New Virginia Curator Credited with Transformation of Brooklyn Zoo," *The Virginian-Pilot*, 19 Aug. 2002, <http://www.virginiazoo.org/pressroom.php?Content=release1> (4 June 2003); Tevah Platt, "An Elegant Affair Celebrates the Magic of Staten Island," *Staten Island Advance*, 16 Feb. 2007, <http://www.silive.com> (19 Feb. 2007); "Zoo's Reptile Wing Receives Award of Merit from New York Construction," *Staten Island Advance*,13 April 2008.

39. Harry Hopkins, *Inventory: An Appraisal of Results of the Works Progress Administration*, 20; "Moses Defends High Park Costs," *New York Times*, 20 Jan. 1940, ProQuest Historical Newspapers.

Bibliography

Archival and Other Primary Sources

Audubon Park Commission Records, Earl K. Long Library, University of New Orleans.

Ben Evans Recreation Program History Collection, Seattle Municipal Archives, Staten Island Zoo Archives.

Brookfield Zoo Archives.

Don Sherwood Parks History Collection, Seattle Municipal Archives.

Ernest Unterman Papers [microform], 1933–1956. Milwaukee Micro Collection 23. University Manuscript Collection. Archives. UWM Libraries. University of Wisconsin–Milwaukee.

Ken Kawata, personal communication, Feb. 2007.

Lucille and William Mann Records, Smithsonian Institution Archives.

Mayor Robert Sidney Maestri Records 1936–1946, New Orleans Public Library

New York City Municipal Archives.

New York City Parks Library Archival Collection.

Park and Recreation Department General Subject File, Dallas Municipal Archives

Ray Pawley, personal communication, 10 May 2008.

Robert Moses papers, Manuscripts and Archives Division. New York Public Library. Astor, Lenox and Tilden Foundations.

St. Louis Zoo Records, 1910–1940, Roll 6, Western Historical Manuscript Collection, University of St. Louis-Missouri.

Wildlife Conservation Society Archives, New York.

Woodland Park Zoo Archives.

Government Documents

The Emergency Work Relief Program of the F.E.R.A., April 1, 1934—July 1, 1935. Washington, D.C.: U.S. Government Printing Office, 1935.

Hopkins, Harry L. *Report on Progress of the Works Program, October 15, 1936.* Washington, D.C.: U.S. Government Printing Office, 1936.

United States. Congress. Senate. Hearing before the Special Committee on Conservation of Wild Life Resources. *Protection and Preservation of the Brown and Grizzly Bears of Alaska*, January 18, 1932.

_____. _____. _____. Subcommittee on the Smithsonian Institution of the Committee on Rules and Administration. *Federal Assistance for Zoos and Aquariums: Hearing before the Subcommittee on the Smithsonian Institution of the Committee on Rules and Administration*, 93rd Cong., 2nd sess., 1974, 3–23

_____. Forest Service. Sale Prospectus, 335,000,000 Cubic Feet National Forest Pulp Timber, West Admirality Island

Unit, Tongass National Forest, Alaska. Washington, 1921.

Books, Articles, & Documentaries

Amaral, Afrânio do. The Anti-Snake-Bite Campaign in Texas and the Sub-Tropical United States." *Bulletin of the Antivenin Institute of America*, October 1927: 77–85.

_____. "The Brazilian Contribution Towards the Improvement of the Specific Snake Bite Treatment," *Proceedings of the New York Pathological Society*, 1923, 89–98.

_____. "Campanhas Anti-Ophidicas" in Memorias do Instituto Butantan, Tomo V, São Paulo Brazil. 1930.

_____. *A General Consideration of Snake Poisoning and Observations on Neotropical Pit-Vipers*. Cambridge, Mass.: Harvard University Press, 1925.

_____. *Serpentes em Crise: á luz de uma legítima defensa no "casa do Butantan."* (São Paulo, 1941.

_____. The Snake Bite Problem in the United States and In Central America. *Bulletin of the Antivenin Institute of America*, Vol. 1, July 1927, No. 2, 31–35.

American Association of Zoological Parks and Aquariums. *Zoological Parks and Aquariums*, 1932.

Austin, William A. *The First Fifty Years: An Informal History of the Detroit Zoological Park and the Detroit Zoological Society*. Detroit: Detroit Zoological Society, 1974.

Baker, Caroline A. A., "The Central Park Zoo: Past and Present." Unpublished manuscript, 1985.

Benchley, Belle J. *My Friends the Apes*. Boston: Little, Brown, 1942.

Berner, Richard C. *Seattle 1921–1940: From Boom to Bust*. Seattle: Charles Press, 1992.

Bingham, Harold C. "Sex Development in Apes." *Comparative Psychology Monographs*, Vol. 5, No. 1, May, 1928: 1–165.

Brazil, Oswaldo Vital. "History of the Primordia of Snake-Bite Accident Serotherapy." *Memorias do Instituto de Butantan*, Vol. 49, No. 1, 1987: 7–20.

Bridges, William. *A Gathering of Animals: An Unconventional History of the New York Zoological Society*. New York: Harper and Row, 1974.

Buickerood, Rick. *The History of the Dallas Zoo Since 1888: A Work-in-Progress about the Oldest Zoo in Texas and in the Southwestern United States*. Dallas: R. Buickerood, 2003.

Burns, Ken. *Huey Long*. *American Collection*. Documentary film, 1987.

Cameron, Jenks. *The Bureau of Biological Survey: Its History, Activities and Organization*. Baltimore: Johns Hopkins University Press, 1929.

Campbell, Sheldon. *Lifeboats to Ararat*. New York: Times Books, 1978.

Caro, Robert. *The Power Broker: Robert Moses and the Fall of New York* (New York: Vintage Books, 1974).

Coe, Jon Charles. "Future Fusion: The Twenty-First Century Zoo." In *Keepers of the Kingdom: The New American Zoo*, ed. Michael Nichols. New York: Thomasson-Grant & Lickle, 1997.

Conant, Roger. *A Field Guide to the Life and Times of Roger Conant*. Provo, UT: Canyonlands Publishing Group, 1997.

_____, and Joseph T. Collins. *Peterson Field Guide to Reptiles and Amphibians Eastern/Central North America*. Boston: Houghton Mifflin, 1998.

Crimmins, Martin Laylor. "Poisonous Snakes and the Anit-Venin Treatmen." *Southern Medical Journal*, July 1929, vol. XXII, no. 7: 603–608.

_____. "Snake-Bites and the Saving of Human Life." *Military Surgeon*, Vol. 74, No. 3, March 1934, pp. 125–32

Crosby, Thomas. *A Zoo for All Seasons: The Smithsonian Animal World*. New York: W.W. Norton, 1979.

Curran, C.H., and Carl Kauffeld. *Snakes and Their Ways*. New York: Harper, 1937.

Cutler, Phoebe. *The Public Landscape of the New Deal*. New Haven, Conn.: Yale University Press, 1985.

Ditmars, Raymond L. *Strange Animals I Have Known*. New York: Harcourt, Brace, 1931.

_____. *Thrills of a Naturalist's Quest*. New York: Macmillan, 1932.

_____, and Lee S. Crandall, *Guide to the New York Zoological Park*. New York: New York Zoological Society, 1939.

Donahue, Jesse, and Erik Trump. *The Politics of Zoos: Exotic Animals and Their Protectors.* Dekalb: Northern Illinois University Press, 2006.

_____. *Political Animals: Public Art in American Zoos and Aquariums.* Lanham, Md.: Lexington Books, 2007.

Eddy, John Whittemore. *Hunting of the Alaska Brown Bear.* New York: G.P. Putman's Sons, 1930.

Ehrlinger, David. *The Cincinnati Zoo and Botanical Garden: From Past to Present.* Cincinnati: Cincinnati Zoo and Botanical Garden, 1993.

Everard, L.C. *Zoological Parks, Aquariums, and Botanical Gardens.* Washington, D.C.: American Association of Museums, 1932.

Forman, L. Ronald. *Audubon Park: An Urban Eden.* New Orleans: Friends of the Zoo, 1985.

Haarmeyer, David, and Elizabeth Larson, "Zoo Inc." In *Private Cures for Public Ills: The Promise of Privatization*, Lawrence Reed, ed. New York: Foundation for Economic Education, 1996.

Hancocks, David. *Animals and Architecture.* New York: Praeger, 1971.

Hanson, Elizabeth. *Animal Attractions: Nature on Display in American Zoos.* Princeton, N.J.: Princeton University Press, 2002.

Heap, Mildred F. *Buffalo Zoological Gardens.* Buffalo, N.Y.: Zoological Society of Buffalo, 1982.

Hoage, R.F., and William A. Deiss, eds. *New Worlds, New Animals: From Menagerie to Zoological Park in the Nineteenth Century.* Baltimore: Johns Hopkins University Press, 1996.

Holzworth, John M. *The Twin Grizzlies of Admiralty Island.* Philadelphia: J.P. Lippincott, 1932.

_____. *The Wild Grizzlies of Alaska.* New York: G.P. Putnam's Sons, 1930.

Hopkins, Harry. *Inventory: An Appraisal of Results of the Works Progress Administration.* Washington, D.C., 1938.

_____. *Spending to Save: The Complete Story of Relief.* New York: Norton, 1936.

Imperato, Pascal James, and Eleanor M. Imperato. *They Married Adventure: The Wandering Lives of Martin & Osa Johnson.* New Brunswick, N.J.: Rutgers University Press, 1992.

Jackson, Dudley. "Treatment of Snake Bite," *Southern Medical Journal,* 1929, July, Vol. XXII, No. 7: 605–607.

Jeffers, H. Paul. *The Napoleon of New York: Mayor Fiorello La Guardia.* New York: John Wiley, 2002.

Katz, Michael. *In the Shadow of the Poorhouse: A Social History of Welfare in America.* New York: Basic Books, 1996.

Kauffeld, Carl. *Snakes: The Keeper and the Kept.* Garden City, N.Y.: Doubleday, 1969.

Kawata, Ken. *New York's Biggest Little Zoo: A History of the Staten Island Zoo.* Dubuque, Iowa: Kendall/Hunt, 2003.

Kisling, Vernon. "Zoological Gardens of the United States." In Vernon Kisling, ed., *Zoo and Aquarium History: Ancient Animal Collections to Zoological Gardens.* Boca Raton, Fla.: CRC Press, 2001.

Larson, Edward J. *Summer for the Gods.* New York: Basic Books, 1997.

Leighninger, Robert J., Jr. *Long-Range Public Investment: The Forgotten Legacy of the New Deal.* Columbia: University of South Carolina Press, 2007.

Lewis, George, and Byron Fish, *I Loved Rogues.* Seattle: Superior, 1978).

Ligibel, Ted. *The Toledo Zoo's First 100 Years: A Century of Adventure.* Virginia Beach, Va.: Donning Co., 1999.

Maple, Terry L., and Erika F. Archibald, *Zoo Man: Inside the Zoo Revolution.* Atlanta: Longstreet, 1993.

Mann, Lucille. *From Jungle to Zoo.* New York: Dodd, Mead, 1934.

Mann, William M. *Ant Hill Odyssey.* Boston: Little, Brown, 1948.

Matthews, Wilber L. *History of the San Antonio Zoo.* San Antonio: San Antonio Zoological Society, 1990.

Mazur, Nicole. *After the Ark: Environmental Policy-Making and the Zoo.* Melbourne: University of Melbourne Press, 2001.

Moses, Robert. *Public Works: A Dangerous Trade.* New York: McGraw-Hill, 1970.

Ohio Writer's Project. *Guide Book: The Cincinnati Zoo.* Cincinnati: Works Progress Administration, 1942.

Parrish, Henry M. *Poisonous Snakebites in the United States*. New York: Vantage Press, 1980.

Perkins, Marlin. *My Wild Kingdom: An Autobiography*. New York: E.P. Dutton, 1982.

Roosevelt, Franklin D. *The Happy Warrior, Alfred E. Smith: A Study of a Public Servant*. Boston: Houghton Mifflin, 1928.

Rosenthal, Mark, Carol Tauber, and Edward Uhlir. *The Ark in the Park: The Story of the Lincoln Park Zoo*. Chicago: University of Illinois Press, 2006.

Ross, Andrea. *Let the Lions Roar! The Evolution of Brookfield Zoo*. Chicago: Chicago Zoological Society, 1997.

Savas, E.S. *Privatization and Public-Private Partnerships*. New York: Chatham, 2000.

Scheier, Joan. *The Central Park Zoo*. Charleston, SC: Arcadia, 2002.

Schwartz, Bonnie Fox. *The Civil Works Administration, 1933–1934: The Business of Emergency Employment in the New Deal*. Princeton, N.J.: Princeton University Press, 1984.

Sclar, Elliott. *You Don't Always Get What You Pay For: The Economics of Privatization*. Ithaca, N.Y.: Cornell University Press, 2000.

Skocpol, Theda, with Margaret Weir, and Anna Shola: *The Politics of Social Policy in the United States*. Princeton, N.J.: Princeton University Press, 1988.

_____. *Protecting Soldiers and Mothers: The Political Origins of Social Policy in the United States*. Cambridge, Mass: Belknap Press, 1992.

Slayton, Robert. *Empire Statesman: The Rise and Redemption of Al Smith*. New York: Free Press, 2001.

Smith, Jason Scott. *Building New Deal Liberalism: The Political Economy of Public Works, 1933–1956*. New York: Cambridge University Press, 2006.

Stillman, Deanne. *Mustang: The Saga of the Wild Horse in the American West*. New York: Houghton Mifflin, 2008.

Townsend, Charles Haskins. *Guide to the New York Aquarium*. New York: New York Zoological Society, 1937.

Tulsa Zoo: The First Fifty Years. Tulsa, Okla.: Tulsa Zoo Development Inc., 1978.

Weiss, Nancy. *Farewell to the Party of Lincoln: Black Politics in the Age of FDR*. Princeton, N.J.: Princeton University Press, 1983.

Wheeler, William Morton. "The Ants of Borneo." *Bulletin of the Museum of Comparative Zoology*, July 1919, Vol. LXIII. No. 3: 43–147.

Wilson, James Q. *Bureaucracy: What Government Agencies Do and Why They Do It*. New York: Basic Books, 1989.

Index

Numbers in ***bold italics*** indicate pages with illustrations.